Advances in Industrial Control

AF789247

Other titles published in this series:

Digital Controller Implementation and Fragility
Robert S.H. Istepanian and James F. Whidborne (Eds.)

Optimisation of Industrial Processes at Supervisory Level
Doris Sáez, Aldo Cipriano and Andrzej W. Ordys

Robust Control of Diesel Ship Propulsion
Nikolaos Xiros

Hydraulic Servo-systems
Mohieddine Jelali and Andreas Kroll

Model-based Fault Diagnosis in Dynamic Systems Using Identification Techniques
Silvio Simani, Cesare Fantuzzi and Ron J. Patton

Strategies for Feedback Linearisation
Freddy Garces, Victor M. Becerra, Chandrasekhar Kambhampati and Kevin Warwick

Robust Autonomous Guidance
Alberto Isidori, Lorenzo Marconi and Andrea Serrani

Dynamic Modelling of Gas Turbines
Gennady G. Kulikov and Haydn A. Thompson (Eds.)

Control of Fuel Cell Power Systems
Jay T. Pukrushpan, Anna G. Stefanopoulou and Huei Peng

Fuzzy Logic, Identification and Predictive Control
Jairo Espinosa, Joos Vandewalle and Vincent Wertz

Optimal Real-time Control of Sewer Networks
Magdalene Marinaki and Markos Papageorgiou

Process Modelling for Control
Benoît Codrons

Computational Intelligence in Time Series Forecasting
Ajoy K. Palit and Dobrivoje Popovic

Modelling and Control of Mini-Flying Machines
Pedro Castillo, Rogelio Lozano and Alejandro Dzul

Ship Motion Control
Tristan Perez

Hard Disk Drive Servo Systems (2nd Ed.)
Ben M. Chen, Tong H. Lee, Kemao Peng and Venkatakrishnan Venkataramanan

Measurement, Control, and Communication Using IEEE 1588
John C. Eidson

Piezoelectric Transducers for Vibration Control and Damping
S.O. Reza Moheimani and Andrew J. Fleming

Manufacturing Systems Control Design
Stjepan Bogdan, Frank L. Lewis, Zdenko Kovačić and José Mireles Jr.

Windup in Control
Peter Hippe

Nonlinear H_2/H_∞ Constrained Feedback Control
Murad Abu-Khalaf, Jie Huang and Frank L. Lewis

Practical Grey-box Process Identification
Torsten Bohlin

Control of Traffic Systems in Buildings
Sandor Markon, Hajime Kita, Hiroshi Kise and Thomas Bartz-Beielstein

Wind Turbine Control Systems
Fernando D. Bianchi, Hernán De Battista and Ricardo J. Mantz

Advanced Fuzzy Logic Technologies in Industrial Applications
Ying Bai, Hanqi Zhuang and Dali Wang (Eds.)

Practical PID Control
Antonio Visioli

(continued after Index)

Carlos Ocampo-Martinez

Model Predictive Control of Wastewater Systems

Springer

Dr. Carlos Ocampo-Martinez
Institut de Robòtica i Informàtica Industrial
Spanish National Research Council (CSIC)
Technical University of Catalonia (UPC)
Parc Tecnològic de Barcelona
C/ Llorens i Artigas 4–6
08028 Barcelona
Spain
cocampo@iri.upc.edu

ISSN 1430-9491
ISBN 978-1-4471-5718-2 ISBN 978-1-84996-353-4 (eBook)
DOI 10.1007/978-1-84996-353-4
Springer London Dordrecht Heidelberg New York

British Library Cataloguing in Publication Data
A catalogue record for this book is available from the British Library

© Springer-Verlag London Limited 2010
Softcover re-print of the Hardcover 1st edition 2010

MATLAB® and Simulink® are registered trademarks of The MathWorks, Inc., 3 Apple Hill Drive, Natick, MA 01760-2098, U.S.A. http://www.mathworks.com

Intel®, Intel Core® and Pentium® are registered trademarks of Intel corporation in the US and other countries.

Apart from any fair dealing for the purposes of research or private study, or criticism or review, as permitted under the Copyright, Designs and Patents Act 1988, this publication may only be reproduced, stored or transmitted, in any form or by any means, with the prior permission in writing of the publishers, or in the case of reprographic reproduction in accordance with the terms of licences issued by the Copyright Licensing Agency. Enquiries concerning reproduction outside those terms should be sent to the publishers.
The use of registered names, trademarks, etc. in this publication does not imply, even in the absence of a specific statement, that such names are exempt from the relevant laws and regulations and therefore free for general use.
The publisher makes no representation, express or implied, with regard to the accuracy of the information contained in this book and cannot accept any legal responsibility or liability for any errors or omissions that may be made.

Cover design: eStudioCalamar, Girona/Berlin

Printed on acid-free paper

Springer is part of Springer Science+Business Media (www.springer.com)

Advances in Industrial Control

Series Editors

Professor Michael J. Grimble, Professor of Industrial Systems and Director
Professor Michael A. Johnson, Professor (Emeritus) of Control Systems and Deputy Director

Industrial Control Centre
Department of Electronic and Electrical Engineering
University of Strathclyde
Graham Hills Building
50 George Street
Glasgow G1 1QE
United Kingdom

Series Advisory Board

Professor E.F. Camacho
Escuela Superior de Ingenieros
Universidad de Sevilla
Camino de los Descubrimientos s/n
41092 Sevilla
Spain

Professor S. Engell
Lehrstuhl für Anlagensteuerungstechnik
Fachbereich Chemietechnik
Universität Dortmund
44221 Dortmund
Germany

Professor G. Goodwin
Department of Electrical and Computer Engineering
The University of Newcastle
Callaghan
NSW 2308
Australia

Professor T.J. Harris
Department of Chemical Engineering
Queen's University
Kingston, Ontario
K7L 3N6
Canada

Professor T.H. Lee
Department of Electrical and Computer Engineering
National University of Singapore
4 Engineering Drive 3
Singapore 117576

Professor (Emeritus) O.P. Malik
Department of Electrical and Computer Engineering
University of Calgary
2500, University Drive, NW
Calgary, Alberta
T2N 1N4
Canada

Professor K.-F. Man
Electronic Engineering Department
City University of Hong Kong
Tat Chee Avenue
Kowloon
Hong Kong

Professor G. Olsson
Department of Industrial Electrical Engineering and Automation
Lund Institute of Technology
Box 118
S-221 00 Lund
Sweden

Professor A. Ray
Department of Mechanical Engineering
Pennsylvania State University
0329 Reber Building
University Park
PA 16802
USA

Professor D.E. Seborg
Chemical Engineering
3335 Engineering II
University of California Santa Barbara
Santa Barbara
CA 93106
USA

Doctor K.K. Tan
Department of Electrical and Computer Engineering
National University of Singapore
4 Engineering Drive 3
Singapore 117576

Professor I. Yamamoto
Department of Mechanical Systems and Environmental Engineering
The University of Kitakyushu
Faculty of Environmental Engineering
1-1, Hibikino,Wakamatsu-ku, Kitakyushu, Fukuoka, 808-0135
Japan

To the health of my mother and my uncle...
and always to the memory of my father

Series Editors' Foreword

The series *Advances in Industrial Control* aims to report and encourage technology transfer in control engineering. The rapid development of control technology has an impact on all areas of the control discipline. New theory, new controllers, actuators, sensors, new industrial processes, computer methods, new applications, new philosophies..., new challenges. Much of this development work resides in industrial reports, feasibility study papers and the reports of advanced collaborative projects. The series offers an opportunity for researchers to present an extended exposition of such new work in all aspects of industrial control for wider and rapid dissemination.

The water and wastewater industry has undergone many changes in recent years. Of particular importance has been a renewed emphasis on improving resource management with tighter regulatory controls setting new targets on pricing, industry efficiency and loss reduction for both water and wastewater with more stringent environmental discharge conditions for wastewater. Meantime, the demand for water and wastewater services grows as the population increases and wishes for improved living conditions involving, among other items, domestic appliances that use water. Consequently, the installed infrastructure of the industry has to be continuously upgraded and extended, and employed more effectively to accommodate the new demands, both in throughput and in meeting the new regulatory conditions. Investment in fixed infrastructure is capital-intensive and slow to come on-stream. One outcome of these changes and demands is that the industry is examining the potential benefits of, and in many cases using, more advanced control systems.

Advanced control in the water and wastewater industry comprises two aspects: technology and control. For *technology*, it is a case of using new distributed digital control hardware and telemetry equipment and moving the industry forward with new working methods and practices based on the new equipment. For *control*, the widespread use of SCADA-type system technology permits the exploitation of more advanced supervisory concepts for total system control and integrated resource management and monitoring. This *Advances in Industrial Control* monograph entitled *Model Predictive Control of Wastewater System* by Carlos Ocampo-Martinez is focused directly on these concerns for the wastewater industry. It poses and gives a possible answer to the question of what advanced control has to offer for the control of a wastewater system with modern technological control capabilities. The system under investigation is part of the Barcelona wastewater system, which is subject to sudden weather-change events within the Mediterranean climate. In a study designed to inform an industrial system operator of the real benefits of the approaches being followed Dr. Ocampo-Martinez ultimately shows the potential percentage reductions in flooding and pollution achievable using the model predictive control (MPC) approach.

The monograph follows a four-part agenda:

Part I Modelling: this section opens with a very instructive Chapter 2 that reports the background to the system infrastructure, and the problems to be found in sewer networks. Detailed modelling of a generic network follows in Chapter 3.

Part II MPC of sewer networks: this part of the monograph takes the reader through three chapters of development work on the application of MPC methods. The novel feature here is the merging of MPC concepts with a hybrid system description of the sewer network. It is instructive to learn how this framework is created and see the results that follow from it.

Part III Fault-tolerant control: more innovative control ideas follow in this part as Dr. Ocampo-Martinez works towards MPC methods that can control the network in the presence of faults on sensors and or actuators in the system.

Part IV Conclusions: in this section, the author looks back at the work accomplished and at the unresolved issues and then looks forward towards new directions for the research.

Clearly, this monograph will be of considerable interest to professionals in the water and wastewater industry. The fact that the work presented is permeated by the realities of modelling and controlling part of the wastewater system of the City of Barcelona will add credibility for this readership. Other engineering professionals who have responsibilities for controlling large-scale geographically distributed networks that have a mixed logical-discrete-continuous technological construction may also be interested to learn what lessons emerge from Dr. Ocampo-Martinez's work in this monograph. MPC is an extremely popular technique for academic researchers of all levels. However, in this monograph, there is a concrete industrial example of MPC extended for use with a hybrid-system process description and this is not so common. Industrial control researchers, academics, graduates and postgraduate students may find the innovative hybrid system development here exciting and worthy of further investigation and research.

Finally, Dr. Ocampo-Martinez's monograph follows previous influential *Advances in Industrial Control* and Springer volumes in this field. In 1999, M.R. Katebi, M.A. Johnson and J. Wilkie, authored the introductory monograph *Control and Instrumentation for Wastewater Treatment Plants* (ISBN 978-1-85233-054-5) that was published as an *Advances in Industrial Control* monograph. In 2002, the Series Editors saw the work of M. Schütze, D. Butler, and B. Beck and encouraged Springer to publish their monograph *Modelling, Simulation and Control of Urban Wastewater Systems* (ISBN 978-1-85233-553-3). Professional contact with M. Marinaki and M. Papageorgiou led to the publication of their monograph *Optimal Real-time Control of Sewer Networks* (ISBN 978-1-85233-894-7, 2005) in the *Advances in Industrial Control* series. With this volume entering the series, the Editors are very pleased to be able to continue the sequence of monographs in this important industrial control-engineering field.

Industrial Control Centre
Glasgow
Scotland, UK
2010

M.J. Grimble
M.A. Johnson

Foreword

Water is one of the fundamental resources of humanity; indeed, it appears that it will be the most influential factor in human development in the near future. As a consequence, the rationalisation of this resource will be a necessary step, and this is where automatic control becomes important. Water flow control is a fundamental tool for preventing unnecessary spillages, flooding and environmental contamination.

This book is therefore timely, as it focuses on the problem of water network control. In particular, it considers the example of sewer network control for the beautiful city of Barcelona – a problem that obviously involves a large-scale system with complex dynamics. This example is developed fully, from modelling to control and finally tolerance against actuator failures.

The book has three main parts. The first presents the sewer network control problem, as well as the state of the art and research topics in this area. To make the book self-contained, the main technical areas are also presented: model predictive control (MPC), hybrid systems and fault-tolerant control tools. A case study where Barcelona's sewer network is fully modelled, calibrated and validated against real rain episodes is then provided.

The second part describes the control strategies used. Here, two approaches are presented: a linear model used in classical MPC and a more general MPC based on a hybrid model formalism denoted HMPC. Advantages of the latter over classical MPC and practical implementation issues are thoroughly discussed.

Finally, the last part of the book considers the practical situation where actuators fail. The previous methodologies are utilised in passive and active approaches. A fault-tolerant hybrid MPC procedure (FTHMPC) is presented and applied to the case study.

This book is a good example of how a thorough treatment of an important application should be performed. It provides an excellent description and addresses the infamous gap between theoretical knowledge and real-world implementation issues. The author has a great deal of experience with these systems, which was gained through his previous doctoral studies and his work with industrial partners. However, the book is not applicable solely to this important application: it could also prove to be very useful for similar hydrological systems.

Buenos Aires, April 2010

Professor Ricardo Sánchez Peña
Principal Investigator, CONICET
Ph.D. Program Director,
Buenos Aires Institute of Technology

Preface

This book is devoted to the modelling and control of sewage systems within the framework of model predictive control (MPC). It also proposes several MPC controller designs for sewer networks, taking into account some of their inherently complex dynamics, the multiple nature of their control objectives, and the performance of the closed-loop scheme when real rain episodes are considered as system disturbances. The incorporation of the closed-loop scheme within a fault-tolerant architecture is studied, and faults in the system actuators (valves and pumping stations) are also considered. The sewage system used as the case study for the book consists of a representative portion of the sewer network of Barcelona (Spain). This portion includes the main phenomena and the most common characteristics and issues that appear in the entire sewer network. Measurements of real rain events and other real data associated with the behaviour of this particular sewer network are available, and the calibrated and validated model of the networked system that follows the corresponding methodology is also included and explained. Results obtained from the simulations associated with this case study highlight the difficulties involved in managing such complex systems.

The book proposes and explores novel research lines in the framework of the real-time control of complex networks related to the urban water cycle. Thus far, several researchers and communities have focused on water management and the dynamical modelling and predictive control of these complex systems as a tool for designing policies aimed at improving the behaviour of water networks, preventing undesirable phenomena caused by different features associated with such networked systems (*e.g.*, weather, topology design, drinking water consumption demands, *etc.*). Additionally, while considering faults in some elements increases the complexity of the problem, detailed research focusing on such adverse system conditions is justified.

Due to the very applied nature of the topics addressed in this book, and the way that different concepts, techniques and methodologies relating to modelling, simulation and control are merged within it, the potential audience of the book consists of academic researchers working in the fields of large-scale networked systems, model predictive control, hybrid systems and fault-tolerant control, researchers associated with water management companies who wish to improve their control algorithms, specialists in sewer network management, people interested in working at the interface between control theory and the real-time implementation of control strategies for complex systems, and postgraduate students interested in research into sewer network modelling, control and fault tolerance.

Barcelona (Spain), April 2010

Carlos Ocampo-Martinez
Institut de Robòtica i Informàtica
Industrial (CSIC-UPC)

Acknowledgements

This book collects results obtained during a pleasant and productive research process. During that period, I met and interacted with many people who each left their particular mark on me; not only in terms of my knowledge of control or mathematics, but also my mind and my memories. Let me express my sincere gratitude to the people and institutions who/which have made it possible prepare this monograph and my Ph.D. thesis (which it builds upon), as well as those who aided in the generation of the main results collected in this book.

This research could not have been done without the invaluable assistance and continuing intellectual and material contributions received from CLABSA (Clavegueram de Barcelona S.A.), which provided the data for the case study, shared its preliminary and expert knowledge about modelling, simulating and controlling sewer networks, and was always willing to advise the research group on the difficult tasks of understanding system behaviour and management criteria.

I also wish to extend my special gratitude to Professor Vicenç Puig, leader of the Advanced Control Systems research group (abbreviated to SAC in Catalan/Spanish), who was one of my Ph.D. advisers and is now the person with whom I share my research on the control of networked systems relating to the urban water cycle. In fact, most of the work described in several chapters of this book resulted from very close collaboration or advice with/from him, which I really appreciate as this unconditional collaboration allowed me to obtain the raw material for this book.

No less important to me and the development of this book is Dr. Ari Ingimundarson and his interesting discussions regarding MPC and hybrid modelling of sewer networks. He was also key to the analysis and interpretation of the main results in some of the chapters. Most of all, he has provided valuable patterns for thinking, learning and acting during my research into the control of complex systems.

Also, and again with the same degree of importance, I wish to thank the various people who were involved in the creation of this book and/or the research it describes. They include Professor Joseba Quevedo, Dra. Gabriela Cembrano, Professor Jan Maciejowski, Professor Alberto Bemporad, among others. Financial support from the CICYT projects ITACA (Integration of Advanced Techniques for Modelling, Control and Supervision Applied to Integral Water Cycle Management – DPI2006-11944) and WATMAN (Analysis and Design of Distributed Optimal Control Strategies Applied to Large-Scale WATer Systems MANagement – DPI2009-13744) is also gratefully acknowledged.

Finally, all of the time that I had to invest into this book meant that, regretfully, I sometimes missed out on the company and affection of my family and close friends. However, they have always been there when I have needed them, which is why I was able to finish this book, and so it is also part of them.

Contents

Part IV Concluding Remarks

List of Tables

List of Figures

Notation

Throughout this book, as a general rule, scalars and vectors are denoted by lower-case letters (*e.g.*, a, x, ...), matrices are denoted by upper-case letters (*e.g.*, A, B, ...), and sets are denoted by double-struck upper-case letters (*e.g.*, $\mathbb{F}$, $\mathbb{G}$, ...). If not otherwise noted, all vectors are column vectors.

$\mathbb{R}$	Set of real numbers
$\mathbb{R}_+$	Set of non-negative real numbers, defined as $\mathbb{R}_+ \triangleq \mathbb{R} \setminus (-\infty, 0]$
$\mathbb{Z}$	Set of integer numbers
$\mathbb{Z}_+$	Set of non-negative integer numbers
$\mathbb{Z}_{\geq c}$	Set defined as $\mathbb{Z}_{\geq c} \triangleq \{k \in \mathbb{Z} \mid k \geq c\}$, for some $c \in \mathbb{Z}$
$\mathbb{Y}^m$	$\underbrace{\mathbb{Y} \times \mathbb{Y} \times \cdots \times \mathbb{Y}}_{m \text{ times}}$
$\lVert q \rVert_p$	Arbitrary Hölder vector p-norm with $1 \leq p \leq \infty$
$\mathbb{A} \subseteq \mathbb{B}$	$\mathbb{A}$ is a subset of $\mathbb{B}$
$\mathbb{A} \subset \mathbb{B}$	$\mathbb{A}$ is a proper subset of $\mathbb{B}$
$\mathbb{A} \cap \mathbb{B}$	Intersection between set $\mathbb{A}$ and $\mathbb{B}$
$\mathbb{A} \oplus \mathbb{B}$	Minkowski sum
$\mathbb{A} \sim \mathbb{B}$	Minkowski (or Pontryagin) difference
A^T	Transpose matrix of $\mathbb{A}$
A^+	Pseudoinverse matrix of $\mathbb{A}$
$\rightarrow$	Mapping
$\mapsto$	Maps to
$\leftrightarrow$	If and only if
H_p	Prediction horizon
H_u	Control horizon
$\mathscr{U}_{H_p}$	Admissible input sequence
$\mathbf{x}_k$	Sequence of states ($\mathbf{x}_k$), control inputs ($\mathbf{u}_k$), logical variables (Δ_k) or auxiliary variables ($\mathbf{z}_k$) over a time horizon m, denoted by $\mathbf{x}_k \triangleq (x_1, x_2, \ldots, x_m)$
$J(\cdot)$	MPC cost or objective function (also denoted by V_{MPC})

w_i	i-th cost function weight
u^*	Optimal value of u
v	Tank volume (state variable)
q_u	Manipulated flow (control input)
d	Rain inflow (measured disturbance)
φ	Ground absorption factor
S	Surface area for a subcatchment
S_w	Wetted surface area of a sewer
β	Volume/flow conversion coefficient
L	Level gauge (limnimeter)
T_i	i-th tank
P	Rain intensity
Δt	Sampling time
$\overline{x}$	Upper bound of the interval where x is defined
$\underline{x}$	Lower bound of the interval where x is defined
$\square\mathbb{A}$	Interval hull of set $\mathbb{A}$ (square box symbol)
diag	Diagonal matrix
min	Minimum
max	Maximum
sat	Saturation function
dzn	Dead-zone function

Acronyms

3-TC	Three-tank catchment
AFC	Actuator fault configuration
AFTHMPC	Active fault-tolerant hybrid model predictive control
AFTMPC	Active fault-tolerant model predictive control
AKF	Adaptive Kalman filter
AS	Automatic supervisor
BTC	Barcelona test catchment
CLABSA	Clavegueram de Barcelona, S.A.
CSO	Combined sewage overflow
CSP	Constraint satisfaction problem
CSS	Combined sewage system
DARE	Discrete-time algebraic Riccati equation
DEDS	Discrete-event dynamical system
FDI	Fault diagnosis and isolation
FTC	Fault-tolerant control
FTHMPC	Fault-tolerant hybrid model predictive control
GPC	Global predictive control
HMPC	Hybrid model predictive control
HYSDEL	Hybrid system description language
ICSP	Interval constraint satisfaction problem
LC	Linear complementarity
LCMPC	Linear constraint model predictive control
LP	Linear program(ming)
LPV	Linear parameter variant
LQR	Linear quadratic regulator
LCRMPC	Linear constrained robust model predictive control
LTI	Linear time invariant
MBPC	Model-based predictive control
MILP	Mixed integer linear program(ming)
MIMO	Multi-input multi-output
MIP	Mixed integer program(ming)

MIQP	Mixed integer quadratic program(ming)
MLD	Mixed logical dynamics
MMPS	Min-max-plus scaling
MPC	Model predictive control
NMPC	Nonlinear model predictive control
mpLP	Multi-parametric linear program(ming)
mpQP	Multi-parametric quadratic program(ming)
NLP	Nonlinear programming algorithms
OOP	Open-loop optimisation problem
PFTHMPC	Passive fault-tolerant hybrid model predictive control
PFTMPC	Passive fault-tolerant model predictive control
PWA	Piecewise affine (*e.g.*, systems, functions, dynamics)
QP	Quadratic program(ming)
RLS	Recursive least squares
RTC	Real-time control
SCADA	Supervisory control and data acquisition
TS	Treated sewage
VFC	Volume/flow conversion

Chapter 1
Introduction

1.1 Motivation

Water – essential to life – will become an increasingly important issue concerning mankind in years to come, because resources of drinking water are relatively scarce on a global scale. Drinking water is the most important renewable natural resource, but it is also the most endangered one. Human activities have led to unsustainable water management and control policies. This drinking water issue is particularly severe along the Mediterranean coast, as a consequence of ongoing climate change. Reports from the IPCC (Intergovernmental Panel on Climate Change, http://www.ipcc.ch/), sponsored by the World Meteorological Organisation and the United Nations, indicate that the availability of hydrological resources in the abovementioned region may decrease by up to 30% in the coming decades.

Problems with drinking water can be associated with human influence on the natural water cycle. Water management has become an increasingly important environmental and socioeconomic subject worldwide. High costs associated with processes such as pumping, transportation, storage, treatment and distribution, and the collection and treatment of urban drainage limit drinking water accessibility for a large portion of the world's population. The processes mentioned before, and others, comprise the *urban water cycle*, which details the long journey of a drop of water from when it is collected from a natural source (*e.g.*, rivers, lakes, dams) until it is returned to its receiving environment within the natural cycle (Mitchell *et al.*, 2001).

When the urban water cycle is understood, it is easier to figure out the difficulties involved in managing it and to infer the most critical problems that need to be solved to do so. Figure 1.1 depicts the urban water cycle, which includes different stages of collection, transport, purification and conditioning for human needs, distribution, consumption, wastewater collection, depuration, and finally reuse or disposal in the natural environment.

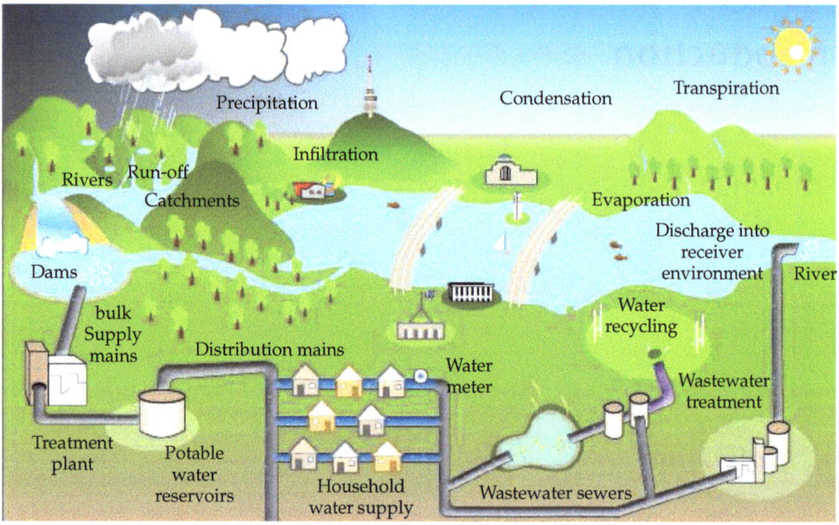

Figure 1.1 Urban water cycle and its main elements and processes

This book focuses on the stage where the sewage produced by homes and businesses is collected and carried to treatment plants in order to avoid polluting the environment. All of the water used in buildings and factories leaves as wasterwater through a set of pipes called *sewer pipes*. Thus, the set of linking pipes and other complementary elements is called the *sewer network*, which is the type of system that is discussed in this book (Figure 1.2).

A sewer network may also incorporate a stormwater system, which collects all of the run-off from rainwater (such as road and roof drainage), and a wastewater treatment system, which treats the sewage and returns it to the natural water cycle as free of pollution as possible. The integration of all of these subsystems increases the complexity of the overall system from the perspectives of management and the potential risks associated with incorrect system operation.

1.1.1 *Sewer Networks as Complex Systems*

According to the previous discussion, sewage systems exhibit some specific characteristics that make them especially challenging from the point of view of analysis and management. These may include many complex features and/or behaviours that can be outlined as follows:

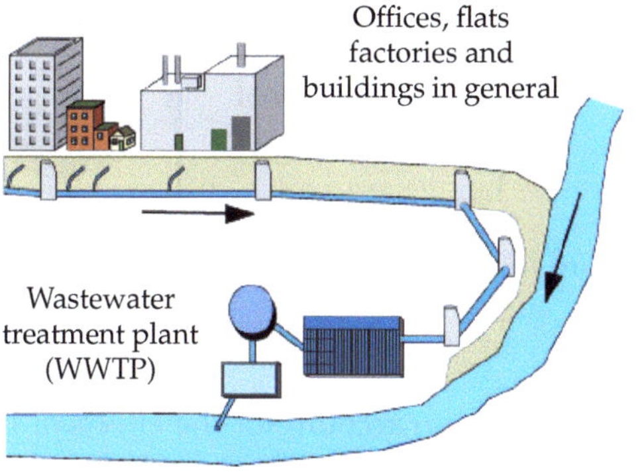

Figure 1.2 Simple sewer network scheme

- Nonlinear dynamics, which can be seen as structural nonlinearities and changes in system parameters according to a given operating point, *e.g.*, in open-flow channel dynamics and in water quality decay models.
- Compositional subsystems with important delays, *e.g.*, in dynamics related to rivers and open-flow channels.
- Compositional subsystems containing both continuous-variable elements (such as pipe flows) and discrete on-off control devices (such as fixed-speed pumps).
- Storage and actuator elements with operating constraints, which are operated within a specific physical range.
- Stochastic disturbances such as rain intensities that affect urban drainage modelling and operation.
- Subsystems with behaviours that are not well understood; *e.g.*, since important physical characteristics such as pipe diameter, roughness and slope change as a function of time, old networks may not behave as expected. Similarly, sewage leakage is an important unknown factor.
- A distributed, large-scale architecture, since water systems generally have hundreds or even thousands of sensors, actuators and local controllers.

All of these features should be taken into account in not only the topological design of the sewer network, but also the definition of an adequate control strategy that fulfils a given set of control objectives. In the case of sewer networks, these objectives are related to environmental protection and the prevention of disasters produced by either incorrect system management or faulty network elements (sensors, actuators or other constitutive elements), among other factors.

There are several advanced modelling and control/management methodologies that can be applied to networked systems that exhibit very complex large-scale dynamics in order to solve the issues described above. Such

Figure 1.3 Compromise between complexity and accuracy for proper system management

techniques can account for the effects of measured disturbances (rain) and consider possible threats to proper system performance from component faults. However, the use of advanced modelling and control methodologies can make the whole closed-loop scheme more complex and significantly increase the computational burden. Therefore, a good compromise between these two aspects is needed in order to achieve practical solutions of the control problem (Figure 1.3).

1.1.2 Model Predictive Control

The analysis of sewer networks poses new challenges to the scientific community, and requires a detailed knowledge of different control methodologies. Such methodologies should handle the effects of rain disturbances in a robust way and should minimise the complexity and computational burden as much as possible. Since there are many sensors and actuators within a sewer network, the system should be governed using a strategy that can handle multi-variable models, compensate for the effects of undesirable dynamics such as delays and dead times, and consider physical constraints and nonlinear behaviours.

Thus, within the field of sewer network control, there is a strategy that is well suited to the particular issues associated with such systems. This strategy is known as *model-based predictive control*, or more simply *model predictive control* (MPC). MPC is more than a control technique: it is a set of control methodologies that use a mathematical model of the considered system to compute the control actions required to minimise a cost function. This function is constructed by merging selected indices related to system performance. MPC is very flexible in terms of its implementation, and can be used in almost all systems as it is configured according to the model of the plant (Camacho and Bordons, 2004).

As will be discussed in Section 2.2.1, MPC has some features that enable it to effectively handle complex systems such as sewer networks. Examples of these features include: compensation for large delays; consideration of the physical constraints of the system; it is a user friendly and intuitive

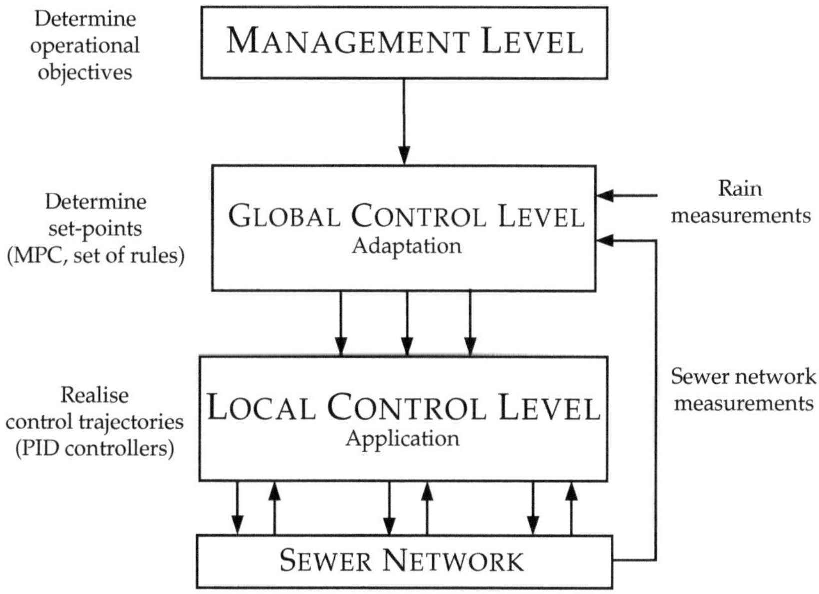

Figure 1.4 Hierarchical structure of an RTC system. Adapted from Schütze *et al.* (2004) and Marinaki and Papageorgiou (2005)

approach (useful for people who do not have a deep knowledge of control); and its ability to handle multi-variable systems. Hence, according to Schütze *et al.* (2004), such controllers a very well suited to the global control of urban drainage systems within a hierarchical control structure (Papageorgiou, 1985; Marinaki and Papageorgiou, 2005). Figure 1.4 shows a conceptual scheme of a hierarchical structure considered in relation to the control of a sewer network.

In Figure 1.4, note that MPC – the global control law – determines the references (set points) for the local controllers placed at different elements of the network. These references are computed according to measurements taken from sensors of various kinds that are distributed around the network (*e.g.*, limnimeters, pluviometers, flowmeters, among others). The management level provides the MPC with its operational objectives: these are reflected in the controller design as the performance indices that are to be enhanced (either minimised or maximised, depending on the case). In the case of urban drainage systems, these indices are usually related to the minimisation or the avoidance (if possible) of flooding and pollution, the minimisation of control energy, *etc.*

1.1.3 *Fault-tolerant Control*

When considering sewer networks, real-time control (RTC) is a custom-designed management architecture for a specific urban sewage system that is activated during a wet-weather event. In some cities around the world, sewer networks use telemetry (rain gauges and water-level sensors in the sewer mains, among other types of sensors) and *telecontrol* in their sewage redirection or sewage retention infrastructures. These elements make it possible to implement active RTC of sewage flows and levels, thus reducing flooding and minimising the risk of discharging pollutants into receiving environments such as seas or rivers.

RTC mechanisms are designed for systems operating under nominal conditions; *i.e.*, with all their elements working properly. However, if a sensor within the telemetry system fails for example, then the designed RTC mechanism should compensate for the resulting misinformation and stop the network from collapsing. Generally, these kinds of faults are caused by extreme meteorological conditions that are typical of the weather in some regions of the world. On the other hand, suppose that a fault which restricts the flow through a network control gate occurs. In a heavy rain episode, this fault could cause sewage to spill onto city streets, resulting in flooding and/or pollution in receiving environments. This situation should be compensated for by an RTC mechanism in order to avoid problems and disasters and thus maintain system performance, perhaps with some allowed degree of degradation.

Therefore, sewer networks need not only a control strategy that is designed to improve system performance but also a set of fault tolerance mechanisms that ensure closed-loop system operation despite the presence of a fault within it. MPC controllers can guarantee a certain level of implicit tolerance due to their inherent capabilities, but system performance is enhanced when additional fault-tolerant policies are included in the RTC architecture.

1.2 Main Objectives of the Book

This book focuses on the modelling and control of sewage systems within the framework of the MPC strategy. Therefore, the main objective of this book is to develop MPC laws for controlling sewer networks that take into account some of the complex dynamics associated with such systems and the multi-objective nature of their management, so that good system performance can be maintained whatever the rain conditions. To this end, the incorporation of the closed-loop system into a fault-tolerant architecture and the consideration of faults at system actuators are also studied. For this specific case, only control gates are considered as actuators. The particular sewage system used as the case study in this book is a representative portion of the sewer network of Barcelona. Measurements of real rain episodes as well as other real data relating to its constitutive elements are available for the whole sewer network. To fulfil this overall objective, a set of particular topics are presented and discussed as follows:

1. The formalisation of sewer network modelling and control in the framework of MPC, including the determination of particular aspects relating to the control strategy, such as cost functions, physical and control problem constraints, tuning methods, *etc.*
2. An analysis of the performance of the sewer network MPC for different controller set-ups with respect to other configurations reported in the literature. This implies the exploration of aspects such as the combination of norms in cost functions, testing different tuning methods and constraint management techniques, *etc.*
3. The use of hybrid systems theory to obtain sewer network models that include the switching dynamics caused by overflows in tanks, weirs and sewer mains. These models are used to design predictive controllers.
4. The exploration of alternative solutions to the problem of the high computational burden of designing MPC controllers for sewer networks based on hybrid models.
5. An analysis of the influence of actuator faults in a closed-loop system governed by a predictive controller which, in turn, is based on either a linear or a hybrid model. Moreover, the determination of the limitations of fault-tolerant control (FTC) schemes under the previously mentioned conditions.
6. Consideration of the inherently hybrid nature of FTC systems, achieved using a hybrid systems modelling, analysis and control methodology.
7. The exploration of numerical techniques for satisfying constraints in order to determine (off-line) the feasibility of fulfilling the control objectives in the presence of actuator faults. This approach to tolerance evaluation avoids the need to solve an optimisation problem in order to check whether the control law can deal with the given actuator fault configuration.

1.3 Outline of the Book

This book is organised as described below.

Chapter 2: Background

This chapter introduces the main ideas behind the different topics considered in this book. The first part focuses on a collection of concepts and definitions associated with sewer networks and presents brief descriptions of their main constitutive elements. The chapter also briefly presents the state of the art in RTC of sewer networks and new research directions in this field. Moreover, concepts and definitions regarding MPC and hybrid system formalisms are outlined. Finally, concepts and methods of implementing fault-tolerance mechanisms are presented, and a brief literature review of this topic is included.

Chapter 3: Principles of the Mathematical Modelling of Sewer Networks

Once the generic structure and the methods used to operate sewer networks have been introduced, a control-oriented modelling methodology for such systems is required. This chapter introduces the modelling principles for sewer networks, which follow the *virtual tanks* approach (Gelormino and Ricker, 1994). In this approach, a network can be considered a set of interconnected tanks that are represented by first-order models which relate inflows and outflows to the tank volume. The calibration technique for the whole sewer network model, based on real data for rain intensities and sewage levels at sewer mains, is explained and discussed. The case study to which the proposed control approaches and other methodologies are applied is also presented and described in detail. The case study used in this book corresponds to a portion of the sewer network under the city of Barcelona. The particular mathematical model is obtained (Ballester *et al.*, 1998) and calibrated using real data from representative rain episodes that occurred in Barcelona during the period 1998–2002.

Chapter 4: Formulating the Model Predictive Control Problem

Based on the system model determined for the case study in Chapter 3, this chapter considers a linear representation of the network; *i.e.*, it ignores some inherent switching dynamics produced by network components such as weirs and discontinuous behaviours related to overflows at sewer mains and/or tanks. The idea is to get an optimisation problem with linear constraints in order to formalise a linear constrained MPC for sewer networks. In this framework, the chapter studies the effect of having different norms in the multi-objective cost function relating to the MPC problem, and it proposes a control tuning approach based on lexicographic programming. Such an approach allows the globally optimal solution to be obtained without needing to perform the tedious procedure of adjusting the weights in a multi-objective cost function.

Chapter 5: MPC Problem Formulation and Hybrid Systems

Limitations regarding the MPC design proposed in Chapter 4 motivated the search for different modelling techniques in order to realise a control-oriented model that includes the inherent switching dynamics of some of the constitutive elements within the sewer network while ensuring the globally optimal solution of the optimisation problem behind the designed MPC controller. Therefore, a modelling methodology for hybrid systems is considered in order to achieve the desirable features discussed before. In this way, a detailed methodology for obtaining a hybrid model that considers the whole sewer network as a compositional hybrid system is proposed. Hence, the hybrid MPC (HMPC) for sewer networks is then put forward and discussed, and the associated mixed-integer programming (MIP) problem is presented. Results obtained by applying the HMPC application to the case study are given, and the main conclusions of this work are drawn.

Chapter 6: Suboptimal Hybrid Model Predictive Control

Results from Chapter 5 show the improvement in system performance achieved when the HMPC is used rather than a system with no control. However, the main drawback of this control approach is the computation time required to solve the associated discrete optimisation problem. Simulations and tests indicate that the MIP problem behind the HMPC design is very random in terms of solution times, since it strongly depends on the system disturbances involved (rain intensities) and the initial condition of the system. Therefore, one way to solve these drawbacks is to relax restrictions associated with the MIP problem in order to reduce the computational burden. This approach can yield suboptimal solutions; *i.e.*, the computational time is reduced but the performance of the solution is degraded. This chapter proposes an approach for relaxing restrictions associated with the MIP problem behind the HMPC design.

The chapter also presents an MPC strategy for systems in mixed logical dynamical (MLD) form, where the number of differences between the sequence of modes of the plant and a reference sequence is limited over the prediction horizon. The aim here is to reduce the number of feasible nodes in the MIP problem, thus reducing the computational burden. The stability of the proposed scheme is proven and the practical issue of how to find the reference sequence is presented and discussed (Ingimundarson *et al.*, 2007). This strategy is applied to the hybrid model of the case study, but with particular considerations relating to its behaviour stated (Ingimundarson *et al.*, 2008).

Additionally, this chapter proposes a control-oriented modelling approach that represents the sewage system with *piecewise linear functions*, following the ideas proposed by Schechter (1987). The purpose of this modelling approach is to reduce the complexity of the MPC problem by avoiding the logical variables introduced when obtaining the MLD form of the system for prediction purposes. Results have shown that MPC controller designs based on the PWLF-based modelling approach yield faster control sequence computation as well as more feasible control sequences with respect to the real-time restriction imposed by the sampling time.

Chapter 7: Model Predictive Control and Fault Tolerance

Faults are highly undesirable events for all control systems. As said before, in the case of a sewer network, a fault can potentially completely halt the global control loop, which could result in severe flooding and increased pollution. MPC controllers, as well as all techniques that employ feedback, have an implicit ability to partially circumvent the effects of faults. Moreover, if the predictive controller governs the closed-loop scheme within a fault-tolerant architecture, faults can be compensated for in an even better way. This chapter takes the definitions and concepts relating to fault-tolerance mechanisms from previous sections and utilises them within the predictive control framework of sewer networks (Ocampo-Martinez *et al.*, 2005). The fault-tolerance capabilities inherent to the MPC strategy are discussed, and the concept of parameterising the system model as a function of the faults is explained using a simple motivational example.

When the prediction model of the MPC controller is obtained while considering the plant to be a hybrid system, the inclusion of fault tolerance leads to the fault-tolerant HMPC (FTHMPC). In this framework, two strategies are discussed: using the natural robustness of MPC towards faults in the plant (passive FTHMPC); and taking into account fault-tolerance mechanisms (active FTHMPC). Finally, this chapter proposes ways of implementing a fault-tolerant architecture for sewer networks that considers faults in the network gates.

Chapter 8: Fault-tolerance Evaluation of Actuator Fault Configurations

As an extension of the study of fault tolerance, Chapter 8 evaluates the fault tolerance of a certain actuator fault configuration (AFC) within the context of a linear constrained predictive/optimal control law. Faults in actuators cause changes in the constraints associated with the control signals, which in turn change the set of feasible solutions of the optimisation problem behind the MPC design. This may lead to the situation where the set of admissible solutions for the control objective is empty. One of the aims of this chapter is to provide methods of computing this set and then evaluating the admissibility of a given AFC. In particular, the admissible solutions set for the predictive control problem that includes the effect of faults can be determined by different approaches. Finally, the proposed method is tested on a reduced expression for the case study, which is enough to show the effectiveness of the proposed approach (Guerra *et al.*, 2007).

Chapter 9: Concluding Remarks

This chapter summarises the main results collected in this book and discusses possible avenues for future research.

Part I
Background and Case Study Modelling

Chapter 2
Background

This chapter briefly reviews the fundamentals of the main topics treated in this book. Three sections cover concepts, definitions and schemes relating to sewer networks, model predictive control (including hybrid models), and fault tolerance mechanisms. Moreover, bibliographical references to relevant scientific contributions in journals and important congress and research reports are given for each topic, and their contents are briefly presented and discussed.

2.1 Sewer Networks: Definitions and Real-time Control

2.1.1 Description and Main Concepts

First of all, this section introduces some important concepts and relevant definitions relating to sewer networks. The basic concepts to define are what a sewer network is and what its objective is. In general, *sewers*[1] are pipelines that transport wastewater and rain drains from city buildings and streets to treatment facilities. Sewers connect these items to the horizontal mains. The sewer mains often connect to larger mains, and these are linked to a wastewater treatment plant. Vertical pipes called *manholes* connect the mains to the surface. Sewers are generally gravity powered, although pumps may be used if necessary.

The main type of wastewater that is collected and transported by a sewer network is generally *sewage*, which is defined as the liquid waste produced by humans. This typically contains washing water, faeces, urine, laundry

[1] The word *sewer* comes from the old French *essouier* (to drain), which in turn derives from the Latin *exaquaria* (*ex-*, meaning "out", and *aquaria*, the feminine of *aquarius*, meaning "pertaining to water".

waste, and other liquid or semi-liquid wastes from households and industry. Such a sewer network is known as a *sanitary sewer network*.[2]

Similarly, there are the *storm sewer mains*, which are large pipes that transport stormwater run-off from streets to natural sewage bodies in order to avoid street flooding. When a given network collects not only sewage from houses and industry but also stormwater run-off, it is called a *unitary network* or a *combined sewage system* (CSS). Such sewer networks were built in many older cities, as a mixed system was cheaper to build, but they encounter problems during heavy rains. Hence, these combined systems were designed to handle storms of a certain size. When the sewer mains were overloaded, the sewage would pass from the sewer system into a nearby body of water through a relief main to prevent flooding in the streets or in houses and buildings. This book considers the case of unitary networks, so all concepts and descriptions presented hereafter relate to such systems.

According to the literature, sewer networks can be considered a collection of elements that each provide a specific function. The discussion below focusses on a few typical elements found in a sewer network, while Figure 2.1 gives an idea of how they are interrelated using a scheme for a very small and simple sewer network. Some of the figures presented here relate to the sewer network of Barcelona, which is described in Chapter 3 and forms the case study of this book.

Transport Links

These elements are used not only to connect different parts of the network but also as storage elements when they become sufficiently extensive (*i.e.*, when the total capacity of these links becomes large enough). Regarding their hydrodynamics, This fact also means that hydrodynamic phenomena relating to the manipulation of the sewer network inflow using throttle gates must be considered when this approach is used. In such cases, the so-called *backwater effect*[3] may occur, which adds an extra degree of complexity when attempting to model and simulate the dynamical behaviours of such elements. Moreover, due to the size of the network, transport delays and other nonlinear dynamics should be taken into account when the dynamics of these elements are described. Within a sewer network, there are a variety of links that vary in size, geometric shape, specific function, *etc.* Figure 2.2 shows a large-diameter sewer main found in a real sewer network.

[2] It is also called a *foul sewer*, especially in the UK.

[3] *Backwater* is water that is held or pushed back; for example by a dam or current.

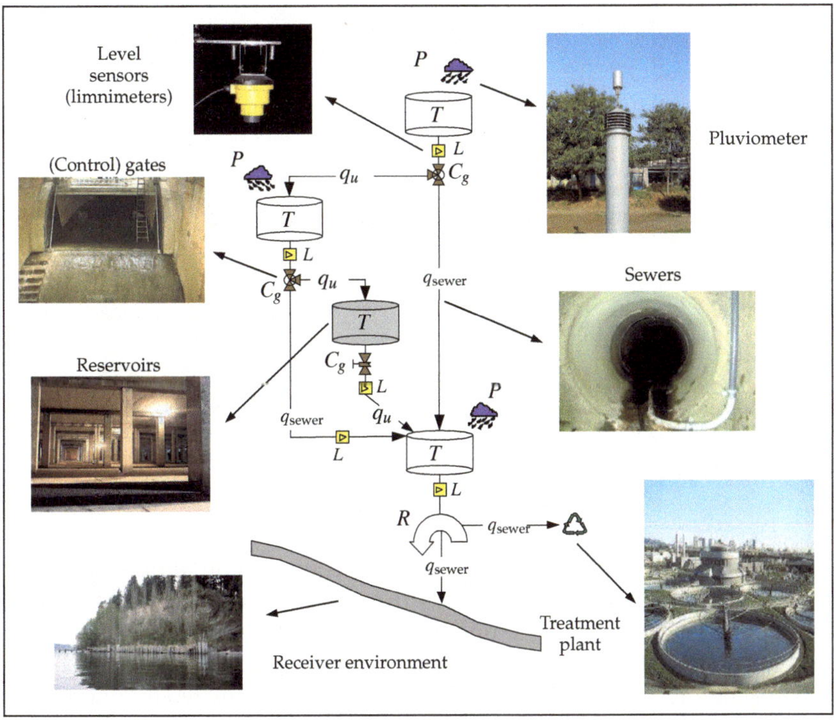

Figure 2.1 Typical components found in the scheme of a simple sewer network

Figure 2.2 Large-diameter sewer. Taken from CLABSA (2007)

Tanks or Reservoirs

These elements are used as dual-function storage devices. Their first function is to ensure that their outflow is laminar, which means that the inflow

Figure 2.3 Inside a retention tank. Taken from CLABSA (2007)

is greater than the outflow. This facilitates easier manipulation of the flows in elements downstream, particularly during heavy rain episodes. Second, these elements have an environmental function in the sense that they retain highly contaminated sewage. This retention prevents the spillage of sewage onto beaches, rivers and ports, thus allowing it to be treated by wastewater treatment plants (WWTPs). On the other hand, the degree of contamination of the sewage retained in a tank decreases due to the sedimentation caused by the retention process.

In terms of phenomenological models, these elements can include overflow capabilities, which means that when the volume of sewage exceeds the maximum capacity of the tank, a new flow can appear. This flow is the sewage that cannot be stored in the tank. However, some models incorporate the manipulation of a redirection gate located in the tank's input as an alternative strategy to cope with overflows. In this case, an overflow is not a nominal mode of operation; it becomes a security mechanism. The maximum capacity of the tank would be a operational constraint on the input gate management policy. The usefulness of each of these approaches depends on the model and the control strategy applied to the sewer system. The inside of a retention tank is shown in Figure 2.3.

Gates

In a sewer network, gates are used as control elements, as they can be used to change the flow downstream. Depending on the actions they perform, gates can be classified as follows:

Figure 2.4 Typical retention gate. Taken from CLABSA (2007)

Redirection gates. These gates are used to change the direction of sewage flow. They can be located before a reservoir or at any position in the network where sewage redirection may be required.

Retention gates. These are used to retain the sewage flow at a certain point in the network. They are generally located at the outputs of reservoirs, where they can be used to control whether sewage is kept in the tank and thus whether it undergoes the beneficial wastewater sedimentation process.

In sewer network control, when a system model based on sewage flows is considered, the control variables may correspond to the manipulated outflows from the network gates. Taking into account the scheme in Figure 1.4, where the global control level computes these outflows, local controllers handle the mechanical actions of the physical gates (actuators) by using these computed outflows as the set-points of those controllers. This procedure avoids the need to consider inherent nonlinearities associated with the dynamic behaviour of the gate. Figure 2.4 shows a large retention gate.

Nodes

According to Marinaki and Papageorgiou (2005), these elements correspond to points where sewage flows are either propagated or merged. Propagation means that the node has one inflow and one outflow, so the objective of this point is – for example – to connect sewer mains with different geometric shapes. On the other hand, merging means that two or more inflows are merged into a greater outflow. Thus, there are two types of nodes:

- Nodes with one inflow and multiple outputs (splitting nodes).
- Nodes with multiple inputs and one output (merging nodes).

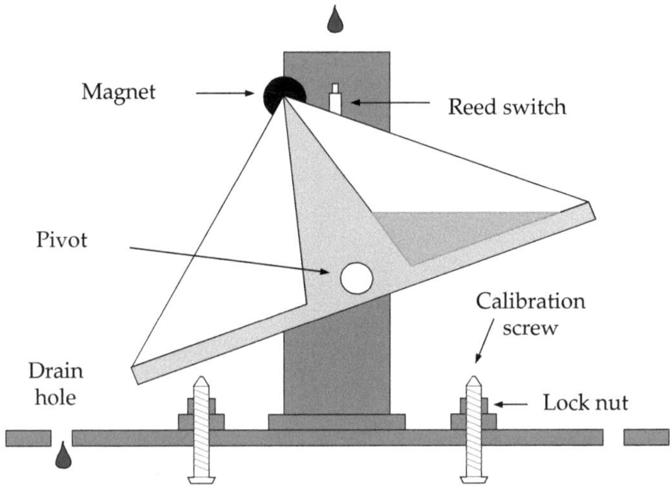

Figure 2.5 Scheme of a tipping bucket rain gauge

Each of these elements inherently has a maximum outflow capacity, which leads to an overflow situation under given conditions. Hence, each node element has a sewage outflow output as well as a possible output for overflow. Such elements are called *weirs*. These exhibit switching behaviour in their dynamics, and such behaviour can be difficult to take into account in many system models.

Instrumentation

In a sewer network, many variables must be measured in order to implement an RTC scheme. The main devices that are used to fulfil this goal are outlined below.

Rain gauges. Rain can be considered to be the main exogenous input to a sewer network. Hence, it is necessary to measure rain intensity in order to compute rain inflow. Rain intensity is measured with a *tipping bucket* rain gauge (see Figure 2.5). This gauge technology uses two small *buckets* mounted on a fulcrum (balanced like a seesaw). The tiny buckets are manufactured with tight tolerances to ensure that they hold an exact amount of precipitation. The tipping bucket assembly is located underneath the rain sewer, which funnels the precipitation to the buckets. As rainfall fills one of the tiny buckets, it overbalances and tips, emptying itself, while the other bucket pivots into place for the next reading. Each tipping event triggers a small switch which activates electronic circuitry that transmits the count to a console indoors; this records the event as a particular amount of rainfall. Once the rain intensity has been determined, the rain inflow can be computed using the procedure proposed and explained in Chapter 3.

Limnimeters. These devices measure sewage levels within the sewer mains. They are placed at strategic points in the network, and the data they provide are related to the sewage volume and flow by means of Manning's formula; see Chapter 3. They are mainly used at locations where the slope of the sewer main allows the sewage to flow due to the effect of gravity.

Velocity sensors. Depending on the geometry and topology of the sewer main considered, the flow information can be inaccurate when it is deduced from level measurements. Therefore, velocity sensors are used to measure the sewage velocity at a specific location in the network. This information enables the sewage flow to be computed more accurately, thus ensuring that situations where the sewage level remains constant in a sewer main with almost no slope and sewage flow do not occur.

Radar. An alternative way to measure the rain intensity is via *weather radar*. Weather radar is an instrument that is used to obtain a detailed description of the spatial and temporal rainfall field. This information is needed to create a hydrological model of a certain region with sufficient resolution. However, such devices are complex instruments. They measure a specific property of raindrops. This property is related to the fraction of the radar beam power that bounces off the target and is detected by the radar. This property, known as the rainfall reflectivity, is indirectly related to the rainfall intensity (through the size distribution of the raindrops). It is also indirectly related to the intensity of the rainfall that reaches the ground (GRAHI, 2007). For more on these instruments and how they are used in sewer network management and control, see Sempere-Torres *et al.* (1999), Bringi and Chandrasekar (2001), Bech *et al.* (2006) and Velasco-Forero *et al.* (2009), among others.

Pumping Stations

Once a rain episode has finished, the part of the sewage volume left in the tanks after the rest of the sewage has exited under the influence of gravity is directed to a WWTP. Two types of elements may be needed to do this: retention gates (discussed earlier) and pumping stations. Pumping stations are needed to remove the sewage from the tanks. Hence, these pumping stations are also manipulated, allowing flow control downstream. These elements usually have complex control strategies that depend on the management policies adopted. Notice that a pumping station can consist of different groups of devices that pump the sewage in a pre-established order, according to sequences of operation. There are many types of pumping stations in sewage systems, such as wet pit pumping stations, pneumatic ejector pumping stations, and dry well pumping stations. Figure 2.6 shows a pumping station of a sewer network.

Figure 2.6 Typical pumping station for a reservoir. Taken from CLABSA (2007)

Treatment Elements

These are basically plants where physicochemical and biological processes are employed to remove organic matter, bacteria, viruses and solids from wastewaters before they are released to rivers, lakes and seas. Such elements are very important components of sewer networks since they help to preserve the ecosystem and maintain the environmental balance within the water cycle. In this context, separating storm sewer mains from waste sewers is a useful strategy, as the huge amount of rainwater inflow generated during a storm can overwhelm WWTPs, resulting in the release of untreated sewage into the environment. Some cities around the world have dealt with this issue by adding large storage tanks or ponds to hold the sewage/rainwater until it can be properly treated. Another way to deal with this is to design a control strategy that prevents all types of pollution and combined sewage overflow (CSO) from the sewer network and thus damage to the environment. Figure 2.7 shows a WWTP located at Baix Llobregat, Barcelona (Spain).

2.1.2 *RTC of Sewage Systems*

This section explores contributions to the literature on the RTC of sewer networks, although it also takes into account aspects of modelling sewage systems due to the close relation between modelling and control for these systems. The RTC of sewage systems plays an important role in meeting increasingly restrictive environmental regulations aimed at reducing the release of untreated wastewater or pollution into the environment. Reducing pollution often requires major investments in infrastructure within city lim-

Figure 2.7 View of the WWTP at Baix Llobregat, Barcelona (Spain). Photograph from http://www.depurbaix.cat/

its, and thus any improvements that can be made to the efficiency of the existing infrastructure (for example through the use of improved control strategies) are of great interest. The advantage of sewer network control has been demonstrated by a number of researchers over the past few years (see, *e.g.*, Gelormino and Ricker, 1994; Pleau *et al.*, 1996; Marinaki, 1999; Pleau *et al.*, 2005; Marinaki and Papageorgiou, 2005). One common control strategy for dealing with urban drainage systems is MPC, as described by Gelormino and Ricker (1994), Pleau *et al.* (2001), Marinaki and Papageorgiou (2005) and Puig *et al.* (2009).

This control methodology is suitable due to the multi-input, multi-output and multi-objective nature of the urban drainage control problem. Additionally, one of the main management goals is to exploit the existing infrastructures of these networked systems, accounting for their size and operational limits. All of these system characteristics are conveniently handled by MPC and its capacity to deal with constraints, multiple control objectives, disturbances, delays, *etc.*

One very important aspect of sewer network management is network modelling. Several modelling approaches have been presented in the literature (see, among others, Ermolin, 1999; Marinaki, 1999; Duchesne *et al.*, 2001; Marinaki and Papageorgiou, 2005; Dellana and West, 2009). Due to the complex nature of the problem, several hydrological models have been proposed (Pleau *et al.*, 1996; Zhu *et al.*, 2001; Vanrolleghem *et al.*, 2005). Sewer networks are systems with complex dynamics, since the sewage flows through sewer mains in open canals. As will be discussed later, flows along open canals can be described by the Saint-Venant partial differential equations, which can be used to perform simulation studies but are highly complex to solve in real time. Control-oriented modelling techniques have also been reported in the literature (see Duchesne *et al.*, 2001; Ocampo-Martinez *et al.*, 2006b). However, when an RTC strategy is implemented, the complexity of the model could imply a high computational burden and thus difficulties in computing a control sequence for a desired performance (Zhu *et al.*, 2001; Marinaki and Papageorgiou, 2005). Additionally, a high computational burden can arise when models with hundreds or even thousands of dynamic variables are considered. Such models of huge dimensions are commonly associated with large-scale systems. The purpose of a given dynamical model is often to perform simulation studies, so they can vary from highly complex partial differential equations to simpler conceptual models.

In an early MPC-related work (Gelormino and Ricker, 1994), a linear model of a sewer network was used for prediction. Linear models that were identified for simulating flows in urban drainage networks using rain measurements as known input were also reported to give good results in Previdi *et al.* (1999). The use of nonlinear models for the predictive control of urban drainage systems has also been reported (see, among others, Ricker and Lee, 1995; Marinaki and Papageorgiou, 1998; Shen *et al.*, 2009; Dellana and West, 2009).

Improvements in prediction achieved by using nonlinear models need to be compared to the uncertainty introduced due to the error in predicting the rain over the horizon. Short-term rain prediction or *nowcasting* is an active field of research (see, among many others, Smith and Austin, 2000; Xu *et al.*, 2005; Chan and Tam, 2005; Suresh, 2007). Using a combination of radar and rain gauge measurements as well as advanced data processing, rain prediction has improved greatly in recent years, and the potential for its use in the predictive control of urban drainage systems has been highlighted in Yuan *et al.* (1999), Elliott and Trowsdale (2007), Thorndahl and Willems (2008) and Villarini *et al.* (2010).

An operational model of an urban drainage system is therefore a set of equations that can rapidly but approximately evaluate the hydrological variables of the network and their responses to gate control actions. In Ricker and Lee (1995), nonlinear MPC was implemented across a large-scale system with 26 states and ten manipulated inputs. It was shown that a complex nonlinear model is always better, but differences from the results obtained with linear MPC may be too small to justify the extra effort required for NMPC. Marinaki and Papageorgiou (1997) state that the use of simpler models for the optimisation-based control of sewer networks is justified because:

- The impact of model inaccuracies is reduced by solving the control problem iteratively and updating inflow predictions and initial conditions
- The details of the local elements and catchments are considered in local control loops.

Regarding control strategies, extensive research has been carried out on the RTC of urban drainage systems. Comprehensive reviews that include a discussion of some existing implementations are given in Schilling *et al.* (1996) (and references therein), Zug *et al.* (2001) and Schütze *et al.* (2008), while practical issues are discussed by Lahoud *et al.* (1998), Schütze *et al.* (2002a) and Akridge and Carty (2006), among others. The common idea here is to use optimisation techniques to improve system performance in terms of avoiding street flooding, preventing CSO discharges, minimising pollution, obtaining uniform utilisation of the sewer system's storage capacity, and, in most cases, minimising running costs, among another objectives (see Ermolin, 1999; Weyand, 2002; Schütze *et al.*, 2002b, 2004). In this way, Gelormino and Ricker (1994) proposed the implementation of MPC of the Seattle urban drainage system. In this work, the authors laid out the fundamental concepts involved in applying these techniques to sewer networks: defining appropriate cost functions, creating and maintaining models, and using prediction to minimise the effect of uncertainty in rain estimation – an aspect that is crucial to the proper operation of such systems in closed loops.

Marinaki and Papageorgiou (1997) proposed the application of optimal control in a previously proposed hierarchical structure (Papageorgiou, 1985). This methodology suggests an RTC structure that combines high efficiency and low implementation cost, and has the following three layers:

- An adaptation layer, where the inflow is predicted (rain) and state estimation is performed in real time
- An optimisation layer, which is responsible for global control and the computation of reference trajectories
- A decentralised control layer, which is responsible for realising the control trajectories.

A similar idea of hierarchical control for RTC can be found in Schütze *et al.* (2004) and Brdys *et al.* (2008). In Marinaki and Papageorgiou (2001), the authors combine the work presented in Marinaki and Papageorgiou (1997) with the receding horizon philosophy; that is, optimal control with a finite time horizon and prediction with a sliding time window. Duchesne *et al.* (2004) implements the global control level introduced in Figure 1.4 within the framework of MPC to minimise the overflow from combined sewer mains during rainfall in the urban area drained by the Marigot interceptor in Laval, Canada. The results showed that allowing surcharged flows in the interceptor during rainfall leads to a significant reduction in overflows.

Although the optimisation methods are applied and (more generally) control procedures are developed in order to determine the optimal (best possible) control action under the given conditions, a suboptimal control decision can be sufficient for RTC so long as it can be ensured that this decision does not cause the system to perform less optimally than in the no-control scenario. However, by applying specific model conditions and MPC strategies, it is possible to ensure that the best solution is obtained.

2.2 MPC and Hybrid Systems

2.2.1 *MPC Strategy Description*

MPC, also referred to as *model-based predictive control*, *receding horizon control*, or *moving horizon optimal control*, is one of the few advanced methodologies that has had a significant impact on industrial control engineering. MPC is applied in the process industry since it can handle multi-variable control problems in a natural form, it can take into account actuator limitations, and it allows constraints to be considered. Predictive control methods are developed around certain common ideas that are discussed in, *e.g.*, Maciejowski (2002), Camacho and Bordons (2004), Goodwin *et al.* (2005) and Rawlings and Mayne (2009), which are basically:

- The explicit use of a model to predict the process output in a time horizon
- The production of a control sequence that minimises a cost (objective) function
- The application of the first control signal from the computed sequence and the displacement of the horizon towards the future.

MPC is a wide field of control methods that are developed around a set of basic elements that the methods have in common. Its parameters can be modified to give different algorithms. These main elements are as follows:

- A prediction model, which should capture all of the process dynamics and can be used to predict the future behaviour of the system output.
- An objective (cost) function, which is, in its general form, the element that penalises deviations of the predicted controller outputs from a reference trajectory. It combines a set of performance indices for the dynamical system considered.
- Control signal computation.

This control strategy presents important advantages over other control methods. Some of these advantages are outlined below (Bordóns, 2000):

- It is very easy to for people lacking a deep knowledge of control to use. Its concepts are quite intuitive and it is relatively easy to tune.
- Can be used to control a wide variety of processes, from those with simple dynamics to systems with big delays, unstable systems, and non-minimum phase systems.
- It is very useful for multi-variable systems.
- It has built-in delay compensation.
- It allows the use of constraints, which can be added during the design process.

However, MPC also has some disadvantages, such as a high computational burden when the control law is calculated. In any case, the main potential problem of this strategy is its strong dependence on the accuracy of the system model. The algorithm for MPC controller design is based on the previous knowledge of the system's behaviour, so its performance is related to the quality of the representation of the plant.

MPC Formulation

In most of the cases presented in the research literature, the MPC formulation is expressed in state space. However, in order to present the strategy in a compact and simple but clear way, let

$$x_{k+1} = g(x_k, u_k) \tag{2.1}$$

be the mapping of states $x_k \in \mathbb{X} \subseteq \mathbb{R}^n$ and control signals $u_k \in \mathbb{U} \subseteq \mathbb{R}^m$ for a given system, where $g : \mathbb{R}^n \times \mathbb{R}^m \to \mathbb{R}^n$ is an arbitrary system state function and $k \in \mathbb{Z}_+$. Let

$$\mathbf{u}_k(x_k) \triangleq \left(u_{0|k}, u_{1|k}, \ldots, u_{H_p-1|k}\right) \in \mathbb{U}^{H_p} \tag{2.2}$$

be the input sequence over a fixed-time prediction horizon H_p. Moreover, the *admissible input sequence* with respect to the state $x_k \in \mathbb{X}$ is defined by

$$\mathscr{U}_{H_p}(x_k) \triangleq \left\{\mathbf{u}_k \in \mathbb{U}^{H_p} | \mathbf{x}_k \in \mathbb{X}^{H_p}\right\}, \tag{2.3}$$

where

$$\mathbf{x}_k(x_k, \mathbf{u}_k) \triangleq \left(x_{1|k}, x_{2|k}, \ldots, x_{H_p|k}\right) \in \mathbb{X}^{H_p} \tag{2.4}$$

corresponds to the state sequence generated by applying input sequence (2.2) to system (2.1) from initial state $x_{0|k} \triangleq x_k$, where x_k is the measured or estimated current state (initial condition). Hence, the receding horizon approach is based on the solution of the open-loop optimisation problem (OOP)

$$\min_{\{\mathbf{u}_k \in \mathscr{U}_{H_p}\}} J(\mathbf{u}_k, x_k, H_p), \tag{2.5a}$$

subject to

$$H_{iq}^{\mathrm{u}} \mathbf{u}_k \leq b_{iq}^{\mathrm{u}}, \tag{2.5b}$$

$$G_{iq} \mathbf{x}_k + H_{iq} \mathbf{u}_k \leq b_{iq}, \tag{2.5c}$$

$$H_{eq}^{\mathrm{u}} \mathbf{u}_k = b_{eq}^{\mathrm{u}}, \tag{2.5d}$$

$$G_{eq} \mathbf{x}_k + H_{eq} \mathbf{u}_k = b_{eq}, \tag{2.5e}$$

where $J(\cdot) : \mathbb{X}_f(H_p) \mapsto \mathbb{R}$ is the cost function with its domain in the *set of feasible states* $\mathbb{X}_f(H_p) \subseteq \mathbb{X}$ (Lazar *et al.*, 2006), H_p denotes the *prediction horizon* or *output horizon*, and G_{iq}, G_{eqe}, H_{iq}, H_{eq}, H^{u}_{iq}, H^{u}_{eq}, b_{iq}, b_{eq}, b^{u}_{iq}, and b^{u}_{eq} are matrices with suitable dimensions. In sequence (2.4), $x_{k+i|k}$ denotes the prediction of the state at time $k+i$ performed at k, starting from $x_{0|k} = x_k$. When $H_p = \infty$, the OOP is called the *infinite horizon problem*; when $H_p \neq \infty$, the OOP is called the *finite horizon problem*. Constraints employed to guarantee the system's stability in a closed loop would be added in (2.5b)–(2.5e). In particular, constraints (2.5d)–(2.5e) are related to static elements where an equality condition must hold.

Assuming that the OOP (2.5) is feasible for $x \in \mathbb{X}$, *i.e.*, $\mathscr{U}_{H_p}(x) \neq \emptyset$, there is an optimal solution given by the sequence

$$\mathbf{u}_k^* \triangleq \left(u^*_{0|k}, u^*_{1|k}, \ldots, u^*_{H_p-1|k} \right) \in \mathscr{U}_{H_p}, \tag{2.6}$$

and then the receding horizon philosophy sets

$$u_{\mathrm{MPC}}(x_k) \triangleq u^*_{0|k}, \tag{2.7}$$

and disregards the computed inputs from $k = 1$ to $k = H_p - 1$, with the whole process repeated at the next time instant $k \in \mathbb{Z}_+$. Expression (2.7) is known in the MPC literature as *the MPC law*.

Remark 2.1. The concept of a *control horizon*, denoted by H_u, was not considered in this statement of the MPC strategy. This concept implies the determination of a number of control actions that may be less than or equal to the number of time predictions (H_p), *i.e.*, $H_u \leq H_p$. The case $H_u > H_p$ is unusual but can be achieved (Maciejowski, 2002). ◊

Summarising, Algorithm 2.1 briefly describes the basic computation procedure for an MPC law.

Algorithm 2.1 Basic procedure for MPC law computation

$k = 0$
loop
  $x_{k+0|k} = x_k$
  $\mathbf{u}_k^*(x_k) \Leftarrow$ solve OOP (2.5)
  Apply only $u_k = u^*_{k+0|k}$
  $k = k+1$
end loop

The way that MPC works is often compared to playing chess. Both follow the philosophy of planning the best sequence of future moves in order to achieve a pre-established objective (in the case of chess: to win the game). At the end, just one movement can be performed.

2.2.2 *Hybrid Systems*

Dynamical systems exhibit several phenomena produced by the interactions of signals of different kinds. In general, systems consist of both continuous and discrete components; the former are typically associated with physical first principles, and the latter with logic devices such as switches, digital circuitry, software code, *etc*. This mixture of logical conditions and continuous dynamics gives rise to a *hybrid system*.

In sewer networks, there are several phenomena (overflows in sewer mains and tanks) and elements in the system (redirection gates and weirs) that exhibit different behaviours depending on the flow/volume present within the network. This fact leads naturally to the use of hybrid models to describe such behaviours. The hybrid models considered here belong to the class of discrete-time linear hybrid systems. The condition of discrete time avoids certain mathematical problems, such as Zeno behaviour (Heymann *et al.*, 2002; Ames and Sastry, 2005), and allows to derive models that can be employed for analysis and to explore optimal/predictive control problems.

Mixed Logical Dynamical Systems

The *mixed logical dynamical* (MLD) modelling framework, introduced by Bemporad and Morari (1999), is one way (among others) of representing hybrid systems that can be described by interdependent physical laws, logical rules, and operating constraints. MLD models have recently been shown to be equivalent to representations of hybrid systems such as *linear complementarity* (LC) systems, *min-max-plus-scaling* (MMPS) systems and *piecewise affine* (PWA) systems (among others) under mild conditions (see Heemels *et al.*, 2001). MLD forms are described by linear dynamical equations that are subject to linear mixed-integer inequalities, *i.e.*, inequalities involving both continuous and binary (or logical, or 0–1) variables. These include physical/discrete states, continuous/integer inputs, and continuous/binary auxiliary variables. The ability to include constraints, constraint prioritisation, and heuristics is a powerful feature of the MLD modelling framework.

The general MLD form can be written as (Bemporad and Morari, 1999)

$$x_{k+1} = Ax_k + B_1 u_k + B_2 \delta_k + B_3 z_k, \tag{2.8a}$$

$$y_k = Cx_k + D_1 u_k + D_2 \delta_k + D_3 z_k, \tag{2.8b}$$

$$E_2 \delta_k + E_3 z_k \leq E_1 u_k + E_4 x_k + E_5, \tag{2.8c}$$

where:

- The x variables are the continuous and binary states

$$x = \begin{bmatrix} x_c \\ x_\ell \end{bmatrix}, \quad x_c \in \mathbb{X} \subseteq \mathbb{R}^{n_c}, \quad x_\ell \in \{0,1\}^{n_\ell} \tag{2.9}$$

- The y variables are the continuous and binary outputs

$$y = \begin{bmatrix} y_c \\ y_\ell \end{bmatrix}, \quad y_c \in \mathbb{Y} \subseteq \mathbb{R}^{p_c}, \quad y_\ell \in \{0,1\}^{p_\ell} \tag{2.10}$$

- The u variables are the continuous and binary inputs

$$u = \begin{bmatrix} u_c \\ u_\ell \end{bmatrix}, \quad u_c \in \mathbb{U} \subseteq \mathbb{R}^{m_c}, \quad u_\ell \in \{0,1\}^{m_\ell} \tag{2.11}$$

- The δ variables are auxiliary Boolean variables with $\delta \in \{0,1\}^{r_\ell}$
- The z variables are auxiliary continuous variables with $z \in \mathbb{R}^{r_c}$.

Note that, by removing (2.8c) and setting δ and z to zero, (2.8a) and (2.8b) reduce to an unconstrained linear discrete-time system in state space. The variables δ and z are introduced when translating logic propositions into linear inequalities. All constraints are collected in the inequality (2.8c).

The transformation of certain hybrid system descriptions into the MLD form requires the application of a given set of rules. To avoid the tedious procedure of deriving the MLD form by hand, a compiler called HYSDEL (HYbrid System DEscription Language) was developed in Torrisi and Bemporad (2004) to generate the matrices A, B_i, C, D_i, and E_i in (2.8).

2.2.3 *MPC Problem and Hybrid Systems*

Different methods for the analysis and design of hybrid systems have been proposed in the literature over the last few years (see, among many others, Bemporad and Morari, 1999; Lygeros *et al.*, 1999; Branicky and Zhang, 2000; Borrelli, 2003; Teel, 2007; Lunze and Lamnabhi-Lagarrigue, 2009). The implementations of these methods directly depend on the hybrid system representation used. One of the most well-studied methods involves the class of optimal controllers, which may use the MLD form in order to compute the corresponding control law according to the system's performance objectives.

The formulation of the optimisation problem in the hybrid MPC (HMPC) framework follows the approach of the standard MPC design (see Maciejowski, 2002). The desired performance indices are expressed as affine functions of the control variables, initial states, and predicted disturbances. However, due to the presence of logical variables, the resulting optimisation problem is a *mixed-integer quadratic* or *linear programming* problem (MIQP or MILP, respectively). The computed control law is referred to as *mixed-integer predictive control* (MIPC).

In general, the structure of the MIPC is defined by the OOP (2.5), with the part related to the switching dynamics added. Hence, the new OOP that considers the hybrid system framework is presented as follows. Assume that the system output should track a reference signal y_r, and that x_r, u_r and z_r are desired references for the states, inputs and auxiliary variables, respectively. For a fixed prediction horizon $H_p \in \mathbb{Z}_{\geq 1}$, the input sequence (2.2) is applied to the system (2.8), resulting in sequence (2.4) and the sequences

$$\Delta_k(x_k, \mathbf{u}_k) \triangleq \left(\delta_{0|k}, \delta_{1|k}, \dots, \delta_{H_p-1|k}\right) \in \{0,1\}^{r_\ell \times H_p}, \tag{2.12}$$

$$\mathbf{z}_k(x_k, \mathbf{u}_k) \triangleq \left(z_{0|k}, z_{1|k}, \dots, zH_p - 1|k\right) \in \mathbb{R}^{r_c \times H_p}, \tag{2.13}$$

under the same conditions as in (2.5).

Hence, the OOP for hybrid systems in MLD form is now defined as

$$\begin{aligned} \min_{\{\mathbf{u}_k \in \mathscr{U}_{H_p}\}, \Delta_k, \mathbf{z}_k} \quad & J(\mathbf{u}_k(x_k), \Delta_k, \mathbf{z}_k, x_k) \triangleq \left\| Q_{x_f}\left(x_{H_p|k} - x_f\right)\right\|_p \\ & + \sum_{i=1}^{H_p-1} \left\|Q_x\left(x_{k+i|k} - x_r\right)\right\|_p + \sum_{i=0}^{H_p-1} \left\|Q_u\left(u_{k+i|k} - u_r\right)\right\|_p \\ & + \sum_{i=0}^{H_p-1} \left(\left\|Q_z\left(z_{k+i|k} - z_r\right)\right\|_p + \left\|Q_y\left(y_{k+i|k} - y_r\right)\right\|_p\right), \end{aligned} \tag{2.14a}$$

subject to

$$x_{k+i+1|k} = Ax_{k+i|k} + B_1 u_{k+i|k} + B_2 \delta_{k+i|k} + B_3 z_{k+i|k}, \tag{2.14b}$$

$$y_{k+i|k} = Cx_{k+i|k} + D_1 u_{k+i|k} + D_2 \delta_{k+i|k} + D_3 z_{k+i|k}, \tag{2.14c}$$

$$E_2 \delta_{k+i|k} + E_3 z_{k+i|k} \leq E_1 u_{k+i|k} + E_4 x_{k+i|k} + E_5, \tag{2.14d}$$

$$x_f = x_{r H_p|k}, \tag{2.14e}$$

for $i = 0, \dots, H_p - 1$, where x_f corresponds to the desired value of the state variable at the end of the prediction horizon, and p is related to the selected norm (1-norm, quadratic norm or infinity norm). Q_{x_f}, Q_x, Q_u, Q_δ, Q_z and Q_y are the weight matrices of suitable dimensions that fulfil the following conditions:

$$\begin{aligned} &Q_{x_f,x,u} = Q^T_{x_f,x,u} \succ 0, \quad Q_{\delta,z,y} = Q^T_{\delta,z,y} \succeq 0 \qquad (p=2), \\ &Q_{x_f}, Q_x, Q_u, Q_\delta, Q_z, Q_y \quad \text{nonsingular} \qquad (p=1, p=\infty). \end{aligned} \tag{2.15}$$

Assuming that MIPC problem (2.14) is feasible for $x \in \mathbb{X}$, there is an optimal solution that is given by the sequence

$$\left(u^*_{0|k}, u^*_{1|k}, \cdots, u^*_{H_p-1|k}, \delta^*_{0|k}, \delta^*_{1_k}, \cdots, \delta^*_{H_p-1|k}, z^*_{0|k}, z^*_{1|k}, \cdots, z^*_{H_p-1|k}\right), \tag{2.16}$$

which, upon applying the receding horizon strategy, yields the MPC law in (2.7). Notice that the described procedure corresponds to the extension of the MPC formulation in Section 2.2.1 to hybrid systems, but that the solution is obtained by solving the OOP (2.14).

Some theoretical aspects of the control of hybrid systems have been significant research topics during the last few years. For instance, notice that H_p should be finite. Infinite horizon formulations are not pragmatic theoretically or in practical implementations. The statement that H_p is as large as possible implies a high number of logical variables in the MIPC problem, which makes a computational treatment almost impossible (Bemporad and Morari, 1999). Thus, the assumption that H_p tends to infinity is even worse in the case of large-scale systems. On the other hand, the constraint $x_f = x_{r H_p|k}$, related to the final state in (2.14), can be relaxed as $x_{H_p|k} \in \mathbb{X}_T \subseteq \mathbb{X}$, where $\mathbb{X}_T$ is defined as the *target state set* (Lazar *et al.*, 2006). According to this assumption, the sequence $\mathcal{U}_{H_p}(x_k)$ in (2.3) is redefined with respect to $\mathbb{X}_T$ as

$$\mathcal{U}_{H_p}(x_k) \triangleq \{\mathbf{u}_k \in \mathbb{U}^{H_p} \,|\, x_k \in \mathbb{X}^{H_p}, x_{H_p|k} \in \mathbb{X}_T\}. \tag{2.17}$$

All of the concepts, formulations and definitions presented so far are used in the following chapters to present MPC formulations for linear and hybrid systems. Chapters 4 and 8 consider the definition of a OOP where the model is purely linear, while Chapters 5, 6 and 7 consider the OOP for a hybrid system.

2.3 Fault-tolerance Mechanisms

The aim of RTC of sewer networks is to improve their dynamical performance during extreme meteorological conditions. Under these conditions, it is very likely that a fault will occur in a constitutive element of the network, which will result in a reduction in control effectiveness, degrading system performance and even causing dangerous situations such as severe flooding or pollution. Therefore, it is very important to incorporate fault tolerance mechanisms that reduce the effects of such faults, thus ensuring that the control objectives are at least partially fulfilled.

Fault-tolerant control (FTC) is a relatively new idea, introduced in the research literature (Patton, 1997), which allows the creation of a control loop that fulfils its objectives (albeit possibly with a degree of degradation) when faults occur in components of the system (instrumentation, actuators and/or a plant). A control loop can be considered fault tolerant if:

- Adaptation strategies for the control law are included in the closed-loop scheme, and/or
- Mechanisms that introduce redundancy in sensors and/or actuators are incorporated.

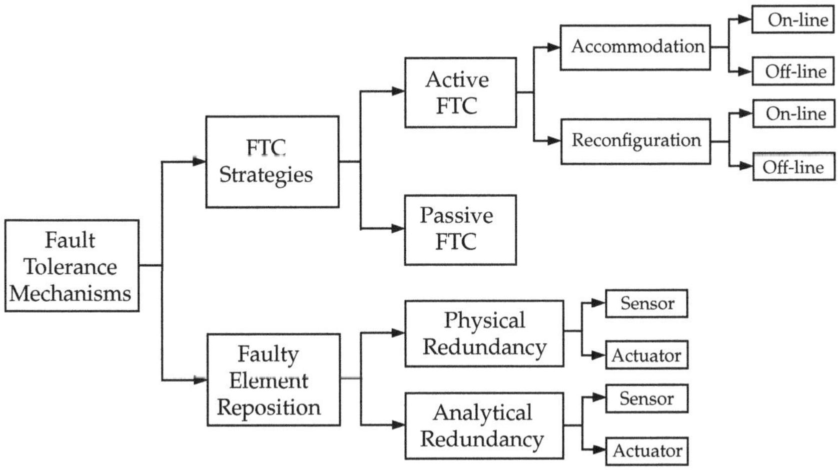

Figure 2.8 Taxonomy of fault-tolerance mechanisms

Figure 2.8 proposes a classification of the fault-tolerance mechanisms considered in this section.

2.3.1 *Fault Tolerance by Adapting the Control Strategy*

The literature considers two main groups of fault-tolerant control strategies: *active* and *passive* techniques. Passive techniques are control laws that account for the occurrence of a fault as a system perturbation. Thus, within certain margins, this type of control law has inherent fault tolerance capabilities that allow the system to cope with the presence of a fault. Complete descriptions of passive FTC techniques can be found in Liang *et al.* (2000), Qu *et al.* (2001), Liao *et al.* (2002), Qu *et al.* (2003), Niemann and Stoustrup (2005), Benosman and Lum (2008) and Steffen *et al.* (2009), among many others.

On the other hand, active FTC techniques adapt the control law based on the information from the FDI block. Using this information, some automatic adjustments are made in order to attempt to achieve the given control objectives.

The scheme of Figure 2.9 proposes a particular architecture of an active FTC loop introduced by Blanke (1999), which contains three design levels: the *control loop* (level 1), the *fault diagnosis and isolation* (FDI) system (level 2) and the *supervisor system* (level 3), which closes the outer loop and adds the fault-tolerance capabilities.

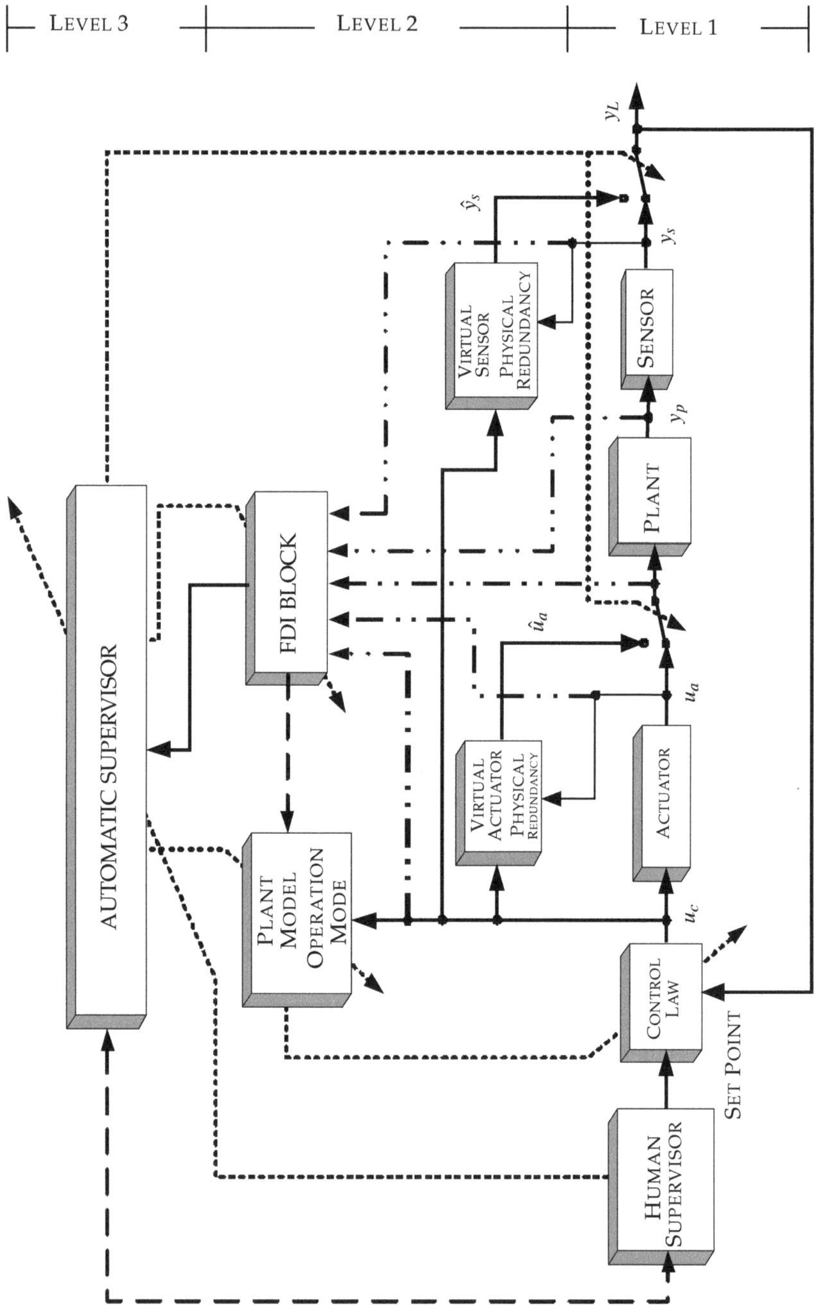

Figure 2.9 Proposed architecture of an FTC system

The feedback control loop shown in Figure 2.9 consists of a *control law*, an *actuator* or a *set of actuators*, the *plant*, and a *sensor* or an *set/array of sensors*. In parallel with the sensor and the actuator blocks, other hardware or software blocks are used to provide *redundancy* during signal measurement and the application of control actions. This redundancy can be introduced in physical form (as redundant sensors or actuators) or in analytical form (by employing mathematical models). Using the input and output signals of sensors, actuators and the plant, the FDI system detects and isolates faults, quantifies the magnitude of the fault, and identifies the faulty elements (if possible). Next, the FDI system sends this information to the *automatic supervisor* (AS), which takes the decisions needed to keep the control loop operative despite the occurrence of the fault.

Note that the AS block is a discrete-event system, while the supervised system is defined in continuous time. Information exchange between these systems is performed through the FDI block. Since the whole system is mixed in nature, its corresponding analysis and design can be done using *hybrid systems theory* (see, among many others, Cassandras *et al.*, 1995; Bemporad and Morari, 1999; Attouche *et al.*, 2001; Morari *et al.*, 2003; Lunze and Lamnabhi-Lagarrigue, 2009), which is an area that is currently being explored and developed in the research literature.

Once the AS block receives the information from the FDI module, it evaluates the admissibility of the system performance, considering the occurrence of the fault. To do this, AS judges whether the control objectives:

- Are fulfilled, allowing for a certain degree of degradation (region of degraded performance), or
- Are not fulfilled, but it is still possible to activate corrective actions (region of unacceptable performance).

Otherwise, the process should be stopped (region of danger). Figure 2.10 shows the abovementioned regions of operation for a two-state system. Chapter 8 presents a methodology that evaluates the admissibility of a given fault configuration.

Accommodation and Reconfiguration Strategies

In order to understand the operation of the different strategies within the active FTC philosophy, the standard feedback control problem is defined by (see Blanke *et al.*, 2006)

$$\langle \mathscr{O}, \mathscr{C}(\theta), \mathscr{U} \rangle, \tag{2.18}$$

where $\mathscr{O}$ is the set of *control objectives*, $\mathscr{C}$ is the set of *system constraints*, θ is the vector of *system parameters* and $\mathscr{U}$ is the *control law*. Hence, the impact of the fault is considered relative to the problem expressed in (2.18), where $\mathscr{C}(\theta)$ indicates how the system constraints depend on the parameters that

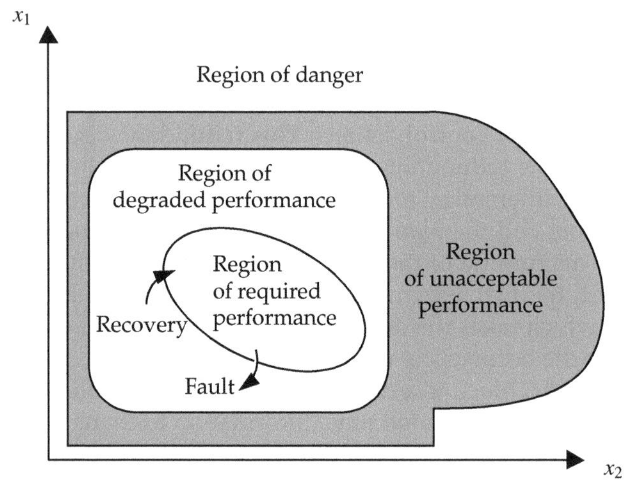

Figure 2.10 Regions of operation based on system performance. Taken from Blanke *et al.* (2006)

may by affected by the faults. The FDI block detects and isolates the fault with or without estimating its magnitude.

Depending on the information provided by the FDI module regarding the magnitude of the fault, two main strategies to adapt the control loop in order to introduce fault tolerance are possible. The first strategy consists of modifying the control law without changing the other elements within the control loop. This is known as *accommodating the system to the effect of the fault*, and it can be done in the case where all changes in system structure and parameters resulting from the fault can be accurately estimated. This concept is introduced formally in Definition 7.1.

The second strategy used to adapt the control loop is based on changing the control law and complementary elements of the closed loop as needed. This is known as *refiguring control due to the presence of a fault*, and it can be applied when no data on fault estimation is available. In this case, faulty components will be unplugged by the FDI block and an attempt will be made to achieve the control objectives using the remaining (fault-free) components. This concept is formally introduced in Definition 7.2.

On-line/Off-line Control Law Adaptation

Once the adaptation approach has been selected for the control law, there are two main ways of implementing this adaptation within the control loop. The basic difference between them is that a pre-computation of the control law parametrised with respect to the faults is performed off-line in one case (off-line adaptation), whereas the control law is recomputed on-line while taking the faults into account in the other case (on-line adaptation). These approaches are described in more depth below:

Off-line adaptation. Also known as adaptation using a pre-calculated controller. In this case, the control law can be written as $\mathscr{U}_f = U(f)$, where f corresponds to the determined fault. Thus, within the FTC architecture, there is a block that determines the mode of operation of the system once the fault occurs, which allows $\mathscr{U}_f$ to be computed (see Figure 2.11 (a)). One possible characterisation of the control laws in this framework according to the nature of the plant (the mathematical model) is given in Theilliol (2003) as follows:

- *Control laws for LTI models*: techniques based on LTI system models, such as *model matching* (Kung, 1992), *model following* (Jiang, 1994), *LQR* and *eigenstructure assigment* (Jiang, 1994; Zhang and Jiang, 2002), among others.
- *Control laws for a family of LTI models*: techniques based on LTI models obtained by linearisation around a set of equilibrium points covering a certain portion of the whole state space. Some examples are *multi-models*, *gain scheduling* and *LPV*, among others.
- *Control laws for nonlinear models*: techniques based on the nonlinear model of the system. In this case, *soft computing* techniques are usually employed to design the controller. Examples of these laws include *fuzzy control*, *neural control* and *neuro-fuzzy control*, among others (Diao and Passino, 2001).

On-line adaptation. Also known as adaptation using a controller computed on-line. In this case, the control law $\mathscr{U}$ is obtained on-line from an estimate of the actual system restrictions $\hat{\mathscr{C}}_f(\hat{\theta}_f)$ once the fault occurs. Figure 2.11 (b) shows the basic operational scheme for this case. There are two ways to estimate the effect of the fault on the system constraints:

- *Off-line estimation*: The effect of the fault on the system constraints is considered off-line. This allows the constraints to be expressed as a function of the fault, and the control law to be changed according to the fault information provided by the FDI module. In this way, the controller is always recomputed while taking into account the effect of the fault on the system constraints. Examples of control techniques from this group are *MPC* (Maciejowski and Ramirez, 1993; Maciejowski and Lemos, 2001) and *static feedback linearisation* (Zhang and Jiang, 2003, 2008).
- *On-line estimation*: The effect of the fault on the system constraints is computed on-line, so the controller changes on-line too. Examples of control techniques from this group are *adaptive control* (Ikeda and Shin, 1995; Diao and Passino, 2002), *dynamic feedback linearisation*

(Zhang and Jiang, 2003, 2008) and *dual predictive control* (Veres and Xia, 1998).

2.3.2 *Fault Tolerance by Repositioning Sensors and/or Actuators*

Serious faults in sensors or actuators break the control loop. In order to keep the system operating, some degree of redundancy should be present so that a new set of sensors (plant inputs) or actuators (plant outputs) can be used. To do this, an accommodation block is implemented to work together with the plant and the other fault-free elements. The main objective is to get a closed loop with almost the same performance as the fault-free closed loop, and thus to attempt to maintain the desired control objectives.

The required redundancy for sensor/actuator fault tolerance can be achieved using either physical redundancy (also called *hardware redundancy*) or analytical redundancy (also known as *software redundancy* or *redundancy by virtual element*).

Fault Tolerance in Sensors

In the case of sensors, the physical redundancy is achieved by having an odd number of measuring elements, and the outputs from these elements are multiplexed in a decision block. Such a block identifies the correct measurement by determining the most common signal value from among the multiplexed signals. On the other hand, the tolerance mechanism utilised with analytical redundancy is to employ an observer, which enables the system measurements to be reconstructed from other existing sensors. Therefore, this technique is also known as a *virtual or software sensor*. The design of a sensor network that takes fault tolerance, system observability, costs and robustness into account is currently an important research subject in the literature (see, *e.g.*, Hoblos *et al.*, 2000; Attouche *et al.*, 2001; Khanna *et al.*, 2004; Kuhn *et al.*, 2006; Michaelides and Panayiotou, 2009). In Staroswiecki *et al.* (2004), sensor network design for fault tolerance estimation is proposed. In this work, aspects like the reliability of a set of sensors, their fault tolerance capabilities and the minimum number of redundant sensors are evaluated. Applications of these mechanisms can be found in aeronautics (Lyshevski *et al.*, 1999; Huo *et al.*, 2001), in AC systems (Bennett *et al.*, 1999), and in wireless network set-ups (Cardei *et al.*, 2007; Chao and Chang, 2008; Alwan and Agarwal, 2009), among many other fields.

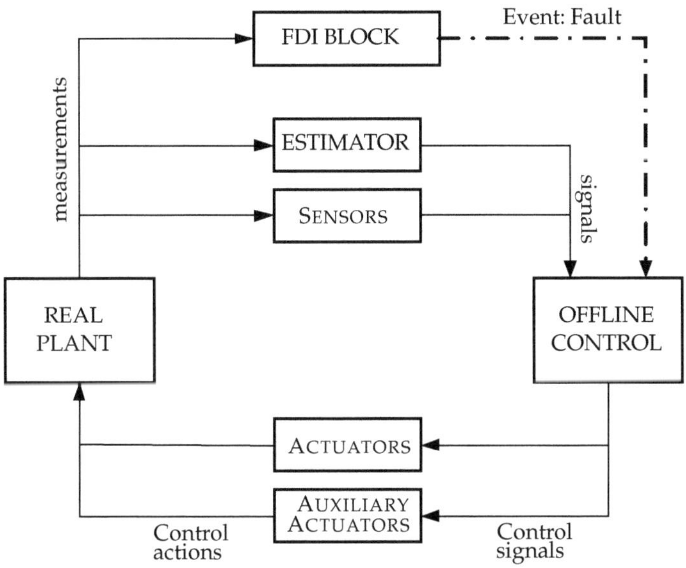

(a)

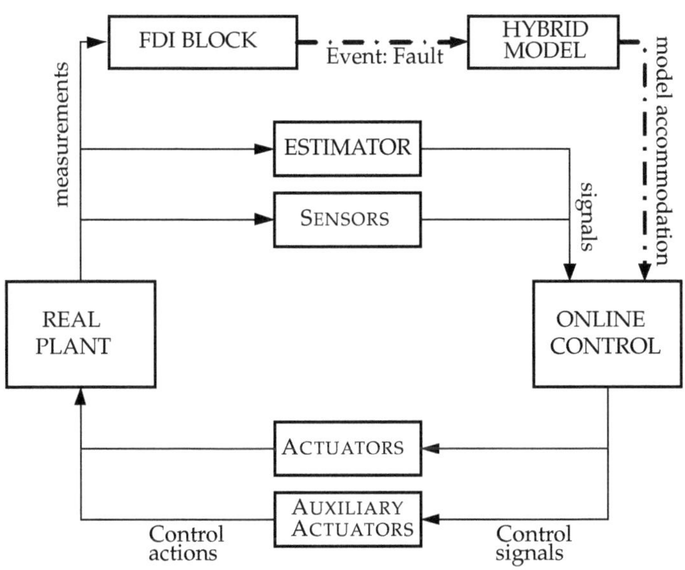

(b)

Figure 2.11 Conceptual schemes for FTC law adaptation: (a) off-line, and (b) on-line

Fault Tolerance in Actuators

Just as for sensors, physical redundancy of actuators means using additional units that can be multiplexed in a decision block by unplugging the faulty actuator and connecting an alternative one with no fault.

On the other hand, in the case of an overactuated system, some degree of physical redundancy already exists. This fact allows to adapt the control law (using either an accommodation or reconfiguration strategy) to find suitable control actions for fault-free actuators. In this way, the control objectives can be fulfilled with an acceptable level of degradation (Dardinier-Maron *et al.*, 1999). Thus, the need to incorporate new hardware into the closed loop is avoided, which makes it cheaper to implement. For instance, in the case of a large-scale water system, where there are thousands of actuators, this approach is suitable for achieving actuator fault tolerance (see Chapters 7 and 8).

The theoretical use of analytical redundancy to achieve actuator fault tolerance was recently proposed. An actuator counterpart to a virtual sensor, known as a *virtual actuator*, was reported in Lunze and Steffen (2003). This proposal was developed further in Lunze and Richter (2006) and Richter *et al.* (2007), while it was applied in Gawthrop and Ballance (2005) and Steffen (2005).

2.4 Summary

This chapter has presented the main aspects of each of the topics treated in this book. The first section encompassed definitions, concepts and discussions taken from the literature regarding sewer networks and their constitutive elements. Moreover, a brief outline of the state of the art in the RTC of networked systems was provided. In the second section, the MPC strategy, hybrid systems, the MPC formulation, and the RTC strategy for sewage systems was presented and discussed.

Finally, the third section collected together the main ideas about existing fault-tolerance mechanisms. In Chapter 3, the sewer network elements presented in Section 2.1 are described using mathematical modelling principles in order to obtain a model of the case study considered in this book. MPC formulations and concepts for linear systems are applied in Chapter 4, while those for hybrid systems are applied in Chapters 5, 6 and 7. Additionally, in Chapter 7 and Chapter 8, the descriptions and definitions relating to fault tolerance and FTC introduced in Section 2.3 are considered and their application discussed.

Chapter 3
Principles of the Mathematical Modelling of Sewer Networks

One of the most important stages in the RTC of sewer networks, and more generally in the control of dynamical systems, is the proper definition of the mathematical model of the system under study. Indeed, this is very important if an *acceptable* performance and satisfactory results are expected when using some control techniques, such as MPC. This chapter focuses on the determination of a control-oriented sewer network model, taking into account the compromise that must be made between accuracy and complexity (Gelormino and Ricker, 1994). Moreover, this chapter describes a case study based on a real sewer network that is a portion of the Barcelona urban drainage system. Using this case study, various control strategies and their associated advantages and problems are presented and discussed in subsequent chapters.

3.1 Fundamentals of the Mathematical Model

Water flow (or sewage) in sewer mains is termed *open channel*; *i.e.*, the flow shares a free surface with the empty space above. The Saint-Venant[1] equations, based on the physical principles of mass conservation and energy, allow to describe accurately the flow in open-flow channels, such as the sewage channels within a network (Mays, 2004). In their general form, these equations are expressed as

[1] Adhémar Jean Claude Barré de Saint-Venant (1797–1886) was a French mechanician and mathematician who developed the equations for one-dimensional, unsteady, open-channel flow in shallow water – also known as the Saint-Venant equations – which are a fundamental set of formulae that are used in modern hydrological engineering.

$$\frac{\partial q_{x,t}}{\partial x} + \frac{\partial A_{x,t}}{\partial t} = 0, \tag{3.1a}$$

$$\frac{\partial q_{x,t}}{\partial t} + \frac{\partial}{\partial x}\left(\frac{q_{x,t}^2}{A_{x,t}}\right) + gA_{x,t}\frac{\partial L_{x,t}}{\partial x} - gA_{x,t}\left(I_0 - I_f\right) = 0, \tag{3.1b}$$

where $q_{x,t}$ is the flow (m^3/s), $A_{x,t}$ is the cross-sectional area of the sewage flow (m^2), t is the time (s), x is the spatial variable measured in the direction and the sense of the water flow (m), g is the acceleration due to gravity (m/s^2), I_0 is the slope of the sewer (dimensionless), I_f is the slope of friction (dimensionless), and $L_{x,t}$ is the water level inside the sewer (m). This pair of partial differential equations constitutes a nonlinear hyperbolic system that lacks an analytical solution for an arbitrary geometry. Notice that these equations get a high level of detail in the description of the system's behaviour. However, such a level of detail is not useful for real-time control implementation, due to the complexity involved in obtaining the solution to (3.1) and thus the high computational cost associated with it (Crossley, 2005).

Alternatively, several modelling techniques have been presented in the literature that deal with the real-time control of sewer networks (see, among many others, Marinaki and Papageorgiou, 1998; Ermolin, 1999; Duchesne *et al.*, 2001; Marinaki and Papageorgiou, 2005). The principles of the modelling used in this chapter are based on the proposal presented in Gelormino and Ricker (1994). There, the sewage system was divided into connected subgroups of sewer mains and treated as interconnected *virtual tanks* (see Figure 3.1).

Definition 3.1 (Virtual Tank). At any given time, let the *virtual tank* be a storage element that represents the total volume of sewage inside the sewer mains associated with a determined subcatchment of a given sewer network. The sewage volume is computed via the mass balance of the stored volume, the inflows and the outflows related to the sewage mains, and the equivalent inflow associated with rainwater.

Using the virtual tank approach and the sewer network elements presented in Section 2.1, a set of compositional elements are introduced and represented mathematically below.

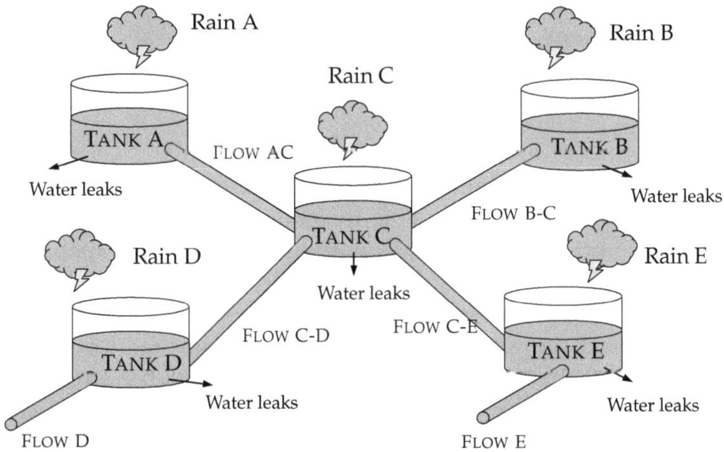

Figure 3.1 Sewer network modelling by means of interconnected virtual tanks

3.1.1 *Virtual and Real Tanks*

These elements are used as storage devices. In the case of a virtual tank, the mass balance of the stored volume, the tank inflows and outflows, and the inflowing rain intensity can be written as a difference equation with a sampling time of Δt:

$$v_{ik+1} = v_{ik} + \Delta t \varphi_i S_i P_{ik} + \Delta t \left(q_{ik}^{\text{in}} - q_{ik}^{\text{out}} \right), \tag{3.2}$$

where v_i corresponds to the volume in the i-th tank at time k (given in cubic meters), φ_i is the *ground absorption coefficient* of the *i-th* catchment, S_i is the corresponding surface area, and P_i is the *rain intensity* at each sample k. Flows q_{ik}^{in} and q_{ik}^{out} are the sum of the inflows and outflows related to the i-th tank, respectively. *Real retention tanks*, which correspond to the reservoirs in the sewer network, are modelled in the same way but without the precipitation term. Tanks are connected by flow paths or links, which represent the main sewage pipes between the tanks. Manipulated variables of the system, which are denoted q_{u_i}, are related to the outflows from the control gates. Tank outflows are assumed to be proportional to the tank volume; that is,

$$q_{ik}^{\text{out}} = \beta_i v_{ik}, \tag{3.3}$$

where β_i (given in s^{-1}) is defined as the *volume/flow conversion* (VFC) coefficient, as suggested in Singh (1988) when using the linear tank model approach. Notice that this relation can be made more accurate (but also more complex) if (3.3) is considered to be nonlinear (thus yielding the nonlinear tank model approach). See Section 3.2 for more details on how to compute this VFC coefficient.

Limits on the volume ranges of real tanks are expressed as

$$0 \leq v_{ik} \leq \overline{v}_i, \tag{3.4}$$

where $\overline{v}_i$ denotes the maximum volume capacity, given in m^3. As this constraint is physical, it is impossible to send more water to a real tank than it can store. Notice that real tanks without an overflow capability are considered. Virtual tanks do not have a physical upper limit on their capacities. When they rise above a pre-established volume, an overflow situation occurs. This fact represents the case where the sewage level in the sewer mains is so high that the sewage can overflow into the streets (flooding). Hence, when the maximum volume of the virtual tanks $\overline{v}$ is reached, any excess volume beyond this maximum amount is redirected to another tank (catchment) within the network or to a receiving environment (as pollution).

3.1.2 Manipulated Gates

Within a sewer network, gates are elements that are used as control devices, as they can change the flow downstream. Depending on the actions they perform, gates can be classified as *redirection gates*, which are used to change the direction of the sewage flow, and *retention gates*, which are used to retain the sewage flow at a certain point in the network (sewer or reservoir). In a real tank, a retention gate is used to control the outflow. Virtual tank outflows cannot be closed, but they can be diverted using redirection gates. Thus, redirection gates divert the flow from the path it normally follows if the gate is closed. This nominal flow is denoted Q_i in the equation below, which expresses the mass conservation relation in the element:

$$q_{ik}^{\text{in}} = Q_{ik} + \sum_j q_{u_i k}^j, \tag{3.5}$$

where j is an index over all manipulated flows $q_{u_i}^j$ from the gate, and q_i^{in} is the flow arriving at the gate. The flow path that Q_i represents is assumed to have a certain capacity, and when this capacity reaches its limit an overflow situation occurs. This flow limit will be denoted $\overline{Q}_i$. When Q_i reaches its maximum capacity, two cases are considered:

1. The water starts to flow out onto the streets, causing a flooding situation
2. The water exits the sewer network and is considered to be lost to a receiving environment.

In the first case, the overflow water either follows the nominal flow path and ends up in the same tank as Q_i, or it is diverted to another virtual tank. Flow into the environment physically represents the situation where the sewage ends up in a river, in the sea, or in another receiver environment.

When this modelling approach is employed, where the inherent nonlinear dynamics of the sewer network are simplified by assuming that only flow rates are manipulated, physical restrictions need to be included as constraints on system variables. For example, the variables $q_{u_i}^j$, which correspond to the desired/manipulated tank outflows, must never be larger than the designed tank outflows (*i.e.*, the tank discharges resulting from the action of gravity). This constraint is expressed by the inequality

$$\sum_j q_{u_i k}^j \leq q_{ik}^{out} = \beta_i v_{ik}. \tag{3.6}$$

Usually, the range of actuation is also limited, so the manipulated variable must fulfil

$$\underline{q}_{u_i}^j \leq q_{u_i k}^j \leq \overline{q}_{u_i}^j, \tag{3.7}$$

where $\underline{q}_{u_i}$ denotes the lower limit of the manipulated flow and $\overline{q}_{u_i}$ denotes its upper limit. When $\underline{q}_{u_i}^j$ equals zero this constraint is convex, but if the lower bound is larger than zero, constraint (3.6) must be included in the range limitation. This leads to the following non-convex inequality:

$$\min(\underline{q}_{u_i}^j, q_{ik}^{out} - \sum_{t \neq j} q_{u_i k}^t) \leq q_{u_i k}^j \leq \overline{q}_{u_i}^j. \tag{3.8}$$

The sum in the expression is calculated for all outflows related to tank i except for j. A further complexity occurs when the control signal is a inflow to a real tank that has hard constraints on its capacity, resulting in a possible situation where the lower bound is also limited by the maximum capacity and the outflow from the real tank.

3.1.3 Weirs (Nodes) and Sewage Pipes

These elements *contribute* switching behaviour to the sewer network, as they describe situations where the sewage flow is constrained due to the pipe capacity, which leads to *jumps* where the flow tries to find an alternative path. According to the discussion in Section 2.1, these elements can be classified as *splitting nodes* and *merging nodes*. The first type can be treated by considering the constant partition of the sewage flow into predefined portions according to topological design characteristics. Merging nodes exhibit switching behaviour. In the case of a set of n inflows q_i, with $i = 1, 2, \ldots, n$, an outflow q_{out} and a maximum outflow capacity $\overline{q}_{out}$, the expressions for this element can be written as follows:

$$q_{in} = \sum_{i=0}^{n} q_i, \tag{3.9a}$$

$$q_{out} = \min\{q_{in}, \overline{q}_{out}\}, \tag{3.9b}$$

$$q_{over} = \max\{0, q_{in} - \overline{q}_{out}\}, \tag{3.9c}$$

where q_{over} corresponds to the node overflow. Notice that these expressions define a nonlinear model of the element, with all the possible implications of this.

Weirs can be thought of as splitting nodes with a maximum capacity in the nominal outflow path that is related to the flow capacity of the output pipe. In the same way, *main sewage pipes* can be seen as weirs with a single inflow. They are used as connection devices between constitutive elements of the network.

3.2 Calibration of Model Parameters

During the determination of control-oriented models for sewage systems, measurements from real network sensors are typically available, and these can be used to estimate the parameters related to compositional elements such as virtual tanks. Measurements of sewage levels in sewer mains are commonly obtained using ultrasonic limnimeters, among other devices. Note that the sewage level is measured rather than its flow rate. This fact avoids the direct contact of the level sensors with the sewage, which prevents problems such as wrong measurements caused by sensor faults. Using these level measurements, inflows and outflows at each tank (either virtual or real) can be estimated assuming steady, uniform flow (Mays, 2004):

$$q = \theta S_w, \tag{3.10}$$

where S_w is the wetted surface, which depends on the cross-sectional area of the sewer A and the water level L within the sewer main. The dependencies of A and L on x and t are omitted for compactness. Moreover, θ is the sewage velocity computed via Manning's formula[2]

$$\theta = \frac{K_n}{n} R_h^{2/3} I_0^{1/2}, \tag{3.11}$$

where K_n is a constant, the value of which depends on the measurement units used in (3.11), I_0 is the slope of the sewer, and n is the Gauckler–

[2] Manning's formula, also known as the Gauckler–Manning formula or the Gauckler–Manning–Strickler formula in Europe, is an empirical formula for open-channel flow, or flow driven by gravity (Goioia and Bombardelli, 2002). It was developed by the French engineer Robert Manning and proposed in 1891 in the *Transactions of the Institution of Civil Engineers of Ireland*.

Manning coefficient, which depends on many factors such as the flow resistance offered by the sewer material, roughness and network sinuosity. The variable R_h is the hydrological radius, defined via the cross-sectional area of the sewage flow and the wetted perimeter[3] p as follows:

$$R_h = \frac{S_w}{p}. \tag{3.12}$$

For a given geometry of the cross-section of the sewer, the wetted perimeter and the hydrological radius can be expressed as a function of the sewage level L. For instance, given a rectangular cross-section of width b, the wetted surface S_w is bL, p is $b+2L$, and the hydrological radius is given by $R_h = \frac{bL}{b+2L}$. Hence, (3.10) is then rewritten as

$$q = \frac{K_n}{n} \frac{(bL)^{5/3}}{(b+2L)^{2/3}} I_0^{1/2}, \tag{3.13}$$

an expression that states the relation between the flow and level of sewage in inflow/outflow sewer mains.

Once sewage level measurements for tank inflows and outflows become available, the dynamical model for virtual and real tanks based on mass conservation can be calibrated. Using the rain intensities P_i and the stated inflows and outflows, and combining (3.2) and (3.3), the mass conservation input/output equation can be expressed as

$$q_{ik+1}^{out} = a\, q_{ik}^{out} + b_1 P_{ik} + b_2 q_{ik}^{in}, \tag{3.14}$$

where $a = (1 - \beta_i \Delta t)$, $b_1 = \beta_i \Delta t \varphi_i S_i$ and $b_2 = \beta_i \Delta t$. Figure 3.2 illustrates this equation and the interactions of all of the parameters and measurements discussed above.

Equation 3.14 is linear in its parameters, which allows them to be estimated using classical parameter estimation methods based on *least-squares algorithms* (Ljung, 1999). Hence, the parameter associated with the ground absorption coefficient is then estimated as

$$\varphi_i = \frac{b_1}{b_2 S_i}, \tag{3.15}$$

and the VFC coefficient is estimated as

$$\beta_i = \frac{b_2}{\Delta t}, \tag{3.16}$$

for the i-th catchment.

Ground absorption and volume/flow conversion coefficients can be estimated on-line at each sampling time using (3.14) and the recursive least-

[3] The *wetted perimeter* is defined as the perimeter of the cross-sectional area that is *wet*.

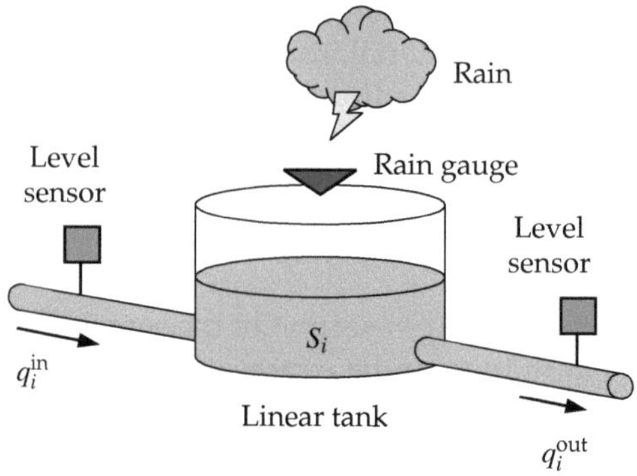

Figure 3.2 Scheme for a single virtual tank showing its associated parameters and measurements

squares (RLS) algorithm (Ljung, 1999). The proper estimation of these parameters ensures that discrepancies between the network sensor information and the prediction obtained from the dynamics in (3.14) are minimised.

Remark 3.1. The level–flow transformation achieved using Manning's formula is very sensitive to topological parameters of the network such as the slope of the sewer main (which modifies the behaviour of θ in (3.11)). If the slope of the sewer main is high, discrepancies between the experimental data and the results obtained from Manning's formula are insignificant. Otherwise, the relation between flow and level is not bijective, so nonlinear behaviours such as the backwater effect lead to significant differences between the experimental data and the results of Manning's formula. This fact led to the consideration of alternative methods of adjusting the relation between measured levels and obtained flows. One way of circumventing possible inaccuracies resulting from the use of Manning's formula is to fit an N-order polynomial to the sewage level measurements. ◊

3.3 Description of the Case Study

3.3.1 *Barcelona's Sewer Network*

The city of Barcelona has a CSS that is approximately 1697 km in length in the municipal area plus 335 km in the metropolitan area, but only 514.43 km are considered to comprise the main sewer network. Its storage capacity is about 3,000,000 m^3, which is three times greater than those for other cities that are comparable in size to Barcelona. The network gathers sewage from about 160,000 points such as connections with buildings (more than 81,500 houses and factories) and the grates at which rainwater enters (known as *scuppers*, which are found in pavements and roads). There are points throughout the network that allow visits. These are termed *wells*, occur every 50 m on average, and number about 30,000 in total.

The main problems with Barcelona's sewer network are caused by three factors: the topology and environment of the city, as well as its population and weather.

Topology and Environment

The topological profile of Barcelona has a significant slope (around 4%) in the zone near to the mountain range that borders the city on one side. This slope decreases towards the Mediterranean Sea (to less than 1%). This aspect causes the rapid accumulation of sewage in zones in the middle of the city and close to the beach when heavy rainstorms occur. Furthermore, the dynamic phenomena associated with its coastal position as well as the occurrence of short but heavy rainstorms and bad marine weather make drainage difficult, considering that these heavy rainstorms can increase the sea level by almost 50 cm (CLABSA, 2007). Similar problems are associated with the drainage of sewage into the rivers Llobregat and Besòs.

Population

One important characteristic of Barcelona is its population. Practically all of Barcelona's 98 km^2 of urban territory is urbanised. Around 1,621,000 inhabitants live in this area,[4] which implies a very high population density (almost 16,000 habitants per km^2). The rapid growth of the city during the twentieth century has left some parts of the sewer network obsolete, so sewage overflows from these areas tend to find their own natural endpoints, thus implying the occurrence of flooding in certain zones downstream.

Weather

The Mediterranean weather of the city and its surroundings can represent another important problem. For example, in some European wetlands, annual precipitation can be greater than 1000 mm (*i.e.*, 1000 $l/m^2/year$), there are on average about 180 days with rain per year, and the intensity of the rain is never greater than 50 mm/h for time intervals of 15 min. However, in the Mediterranean area, and particularly in Barcelona, the annual precipitation is approximately 600 mm (600 $l/m^2/year$) and there are relatively

[4] According to an official report from the Spanish Institute of Statistics (see http://www.ine.es/), dated January 1st, 2009.

few rainy days per year (between 40 and 70), but rain intensities can reach 150 mm/h for periods of 15 min. This implies that some heavy rainstorms – which are typical of Mediterranean weather – can produce a quarter of the annual precipitation in just half an hour, thus representing a headache for those attempting to manage Barcelona's sewer network.

Moreover, it has been shown that the urban environment affects the local climate, which implies a correlation between the second and the third problems. The thermal difference between Barcelona and its surroundings can reach 3°C or 4°C. This phenomenon may encourage rainstorms in terms of both number and intensity.

Sewer Network Management

Clavegueram de Barcelona, S.A. (CLABSA) is the company in charge of sewage system management in Barcelona. A remote control system has been in operation since 1994, and this includes sensors, regulators, remote stations, communications, and the CLABSA Control Centre. In terms of regulators, the urban drainage system currently contains 21 pumping stations, 36 gates, 10 valves and 8 detention tanks, which are regulated in order to prevent flooding and CSO. The remote control system is equipped with 56 remote stations, including 23 rain gauges and 136 water-level sensors, which provide real-time information on rainfall passing into and water levels in the sewage system. All of this information is centralised at the CLABSA Control Centre through a supervisory control and data acquisition (SCADA) system. The regulated elements (pumps, gates and retention tanks) are currently controlled locally; *i.e.*, they are handled from the remote control centre based on the measurements of sensors connected only to the local station (CLABSA, 2007).

3.3.2 Barcelona Test Catchment

This book considers a portion of the whole sewer network of Barcelona that is representative, in that it exhibits the main phenomena of the most common characteristics found in the entire network. This representative portion is used for the case study of this book; a calibrated and validated model of the system that follows the methodology explained in Section 3.2 is also available as rain-gauge data for an interval of several years.

The Barcelona test catchment (BTC) considered in this book covers a surface area of 22.6 km^2 and includes typical elements of the whole network. Due to the size of the catchment area, the rain intensity varies depending on the location of the rain gauge. Figure 3.3 shows the catchment area on a map of Barcelona.

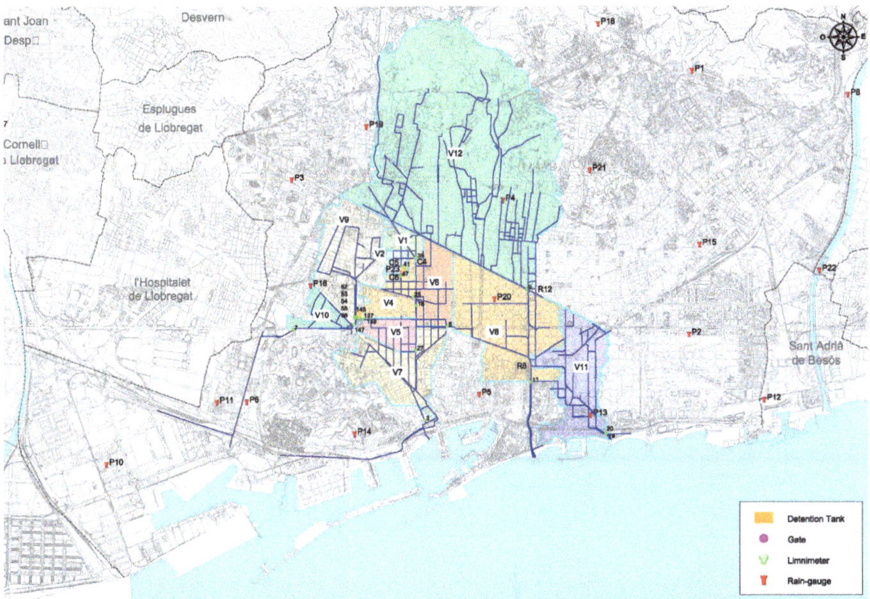

Figure 3.3 Test catchment area on a map of Barcelona. Courtesy of CLABSA

The labels V_i in the figure indicate the different subcatchments that are considered in this book using the notation T_i. Notice that the case study corresponds to an important part of the network, and it is completely representative of the whole sewage system. On the other hand, the equivalent system that utilises the virtual tank approach (Section 3.1) is presented in Figure 3.4 (Ballester *et al.*, 1998).

The BTC has one retention gate associated with a real tank, three redirection gates and one retention gate, eleven subcatchments that define an equal number of virtual tanks, several sewage level gauges (limnimeters), and two WWTPs. There are only five rain gauges, since some virtual tanks share the same rain sensor. These sensors count the number of tipping events in 5 min (the sampling time of this network), and such values are multiplied by 1.2 mm/h in order to obtain the rain intensity P in units of m/s at each sampling time (after the appropriate unit conversion). The difference between the rain inflows for virtual tanks that share a sensor arises from the surface area S_i and the ground absorption coefficient φ_i in (3.15) of the i-th subcatchment. The real tank corresponds to the Escola Industrial reservoir, which is located under a soccer field at the Industrial School of Barcelona (see Figure 3.5). It has a rectanglar geometry of 94×54 m, a medium depth of 7 m and a maximum sewage capacity of $35{,}000\,\text{m}^3$ (CLABSA, 2007).

The related system model has twelve state variables corresponding to the volumes in the twelve tanks (one real, eleven virtual), four control inputs corresponding to the manipulated flows, and five measured distur-

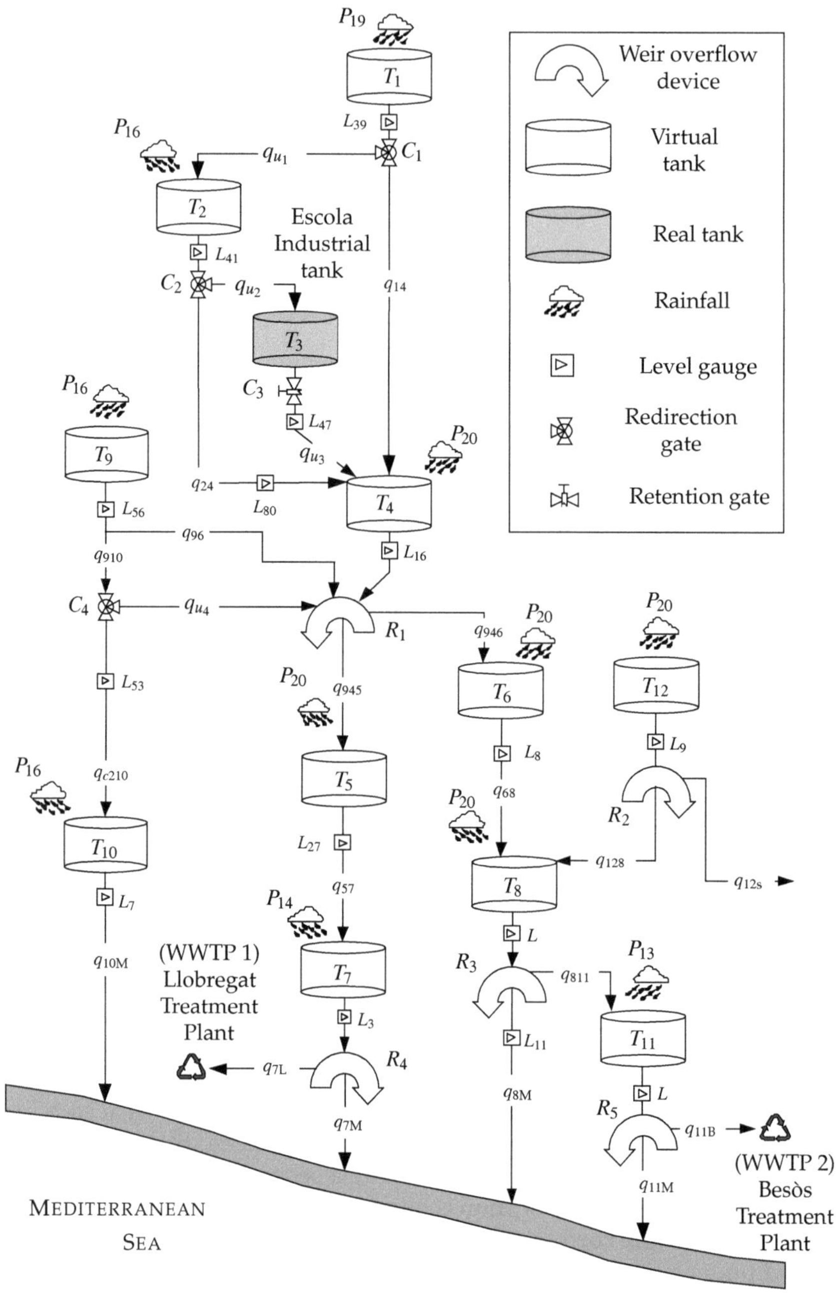

Figure 3.4 Scheme of the Barcelona test catchment

Figure 3.5 Retention tank located at the Escola Industrial de Barcelona

bances corresponding to the measurements of rain precipitation for the virtual tanks. Two WWTPs can be used to treat the sewage before it is released to the receiving environment. Figure 3.4 shows the *free* flows to the environment as pollution ($q_{10\text{M}}$, $q_{7\text{M}}$, $q_{8\text{M}}$ and $q_{11\text{M}}$ to the Mediterranean Sea and q_{12s} to the other catchment), as well as the flows to the treatment plants ($Q_{7\text{L}}$ and $Q_{11\text{B}}$). The variables d_i for $i \in [1,12], i \in \mathbb{Z}, i \neq 3$ are related to rain inflows as a function of the rain intensities P_{13}, P_{14}, P_{16}, P_{19} and P_{20}, depending on the case. The four manipulated flows, denoted q_{u_i}, have maximum flow capacities of $11, 25, 7$ and $29.3\ \text{m}^3/\text{s}$, respectively. Those amounts are fixed, so they become physical constraints of the system.

Regarding the calibration of this particular BTC model (according to the discussion in Section 3.2), Figures 3.6 (a) and 3.6 (b) present comparisons between real levels (from real data) and predicted levels (using the modelling principles described in Section 3.1) in the output sewer mains of virtual tanks T_1 and T_2, respectively. It is apparent that an *acceptable* fit is obtained with the prediction model (and hence the computed parameters are acceptable too) using the abovementioned calibration strategy.

Tables 3.1 and 3.2 describe the case study variables and provide the values of the parameters obtained using the discussed calibration strategy. In Table 3.1 (and also in Figure 3.4), T_i for $i \in [1,12], i \in \mathbb{Z}, i \neq 3$ denotes the i-th subcatchment associated with a virtual tank, while T_3 denotes the real tank. In Table 3.2, $\overline{q}$ denotes the maximum flow capacity for the corresponding sewer main.

Weirs (R_i) can be considered nodes where the path followed by the sewage depends on the flow capacity of the sewer located immediately downstream. The presence of these elements within the network results in the addition of nonlinear expressions to the system model due to their nature and dynamics. This fact motivates the use of control-oriented modelling

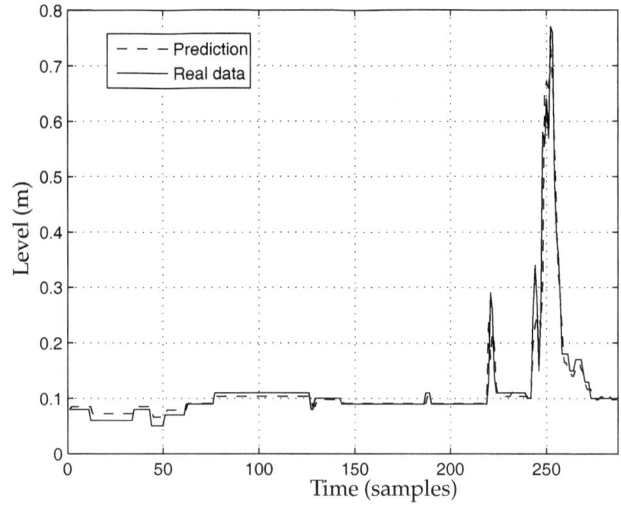

(a)

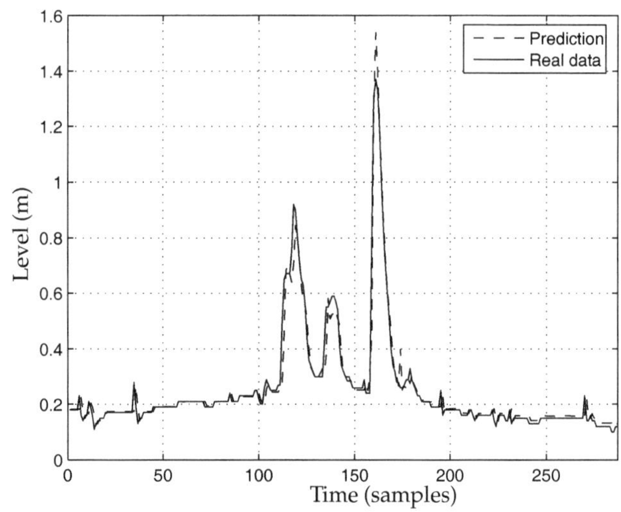

(b)

Figure 3.6 Model calibration results obtained using the strategy given in Section 3.2: (a) output sewage level in T_1, and (b) output sewage level in T_2

methodologies that include such switching dynamics, and, if possible, allows the design of a linear predictive controller and all of its implications in

Table 3.1 Parameter values for the subcatchments of the BTC

Tank	S (m^2)	φ_i	β_i (s^{-1})	$\overline{v}_i$ (m^3)
T_1	323576	1.03	7.1×10^{-4}	16901
T_2	164869	10.4	5.8×10^{-4}	43000
T_3	5076		2.0×10^{-4}	35000
T_4	754131	0.48	1.0×10^{-3}	26659
T_5	489892	1.93	1.2×10^{-4}	27854
T_6	925437	0.51	5.4×10^{-4}	26659
T_7	1570753	1.30	3.5×10^{-4}	79229
T_8	2943140	0.16	5.4×10^{-4}	87407
T_9	1823194	0.49	1.3×10^{-4}	91988
T_{10}	385274	5.40	4.1×10^{-4}	175220
T_{12}	1913067	1.00	5.0×10^{-4}	91442
T_{12}	11345595	1.00	5.0×10^{-4}	293248

Table 3.2 Maximum flows through the sewer mains of the BTC

Sewer	$\overline{q}$ (m^3/s)	Sewer	$\overline{q}$ (m^3/s)
q_{14}	9.14	q_{128}	63.40
q_{24}	3.40	q_{57}	14.96
q_{96}	10.00	q_{68}	7.70
q_{c210}	32.80	q_{12s}	60.00
q_{945}	13.36	q_{811}	30.00
q_{910}	24.00	q_{7L}	7.30
q_{946}	24.60	q_{11B}	9.00

terms of the convexity and globality of the optimal solutions of the optimisation problems associated with MPC strategies.

3.3.3 *Rain Episodes*

The rain episodes used in the BTC simulation and for the design of control strategies are based on real rain-gauge data obtained within the city of Barcelona on certain dates (given as year-month-day), as presented in Table 3.3. These episodes were selected to represent the meteorological behaviour of Barcelona; *i.e.*, they contain representative meteorological phenomena for the city. The table also shows the maximum *return rate*[5] observed among all five rain gauges for each episode. In the third column of the table, the return rate for the whole of Barcelona is shown. This number is lower because it includes a total of 20 rain gauges. Note that one of the rainstorms had a re-

[5] The return rate or return period is defined as the average interval of time within which a hydrological event of a given magnitude is expected to be equalled or exceeded exactly once. In general, this amount is given in years.

Table 3.3 Data for rain episodes

Rain episode	Maximum return rate (years)	Average return rate (years)
99-09-14	16.3	4.3
02-07-31	8.3	1.0
02-10-09	2.8	0.6
99-09-03	1.8	0.6
99-10-17	1.2	0.7
00-09-28	1.1	0.4
98-10-05	1.4	0.2
98-09-25	0.6	0.3
98-10-18	0.4	0.1
00-09-19	0.3	0.2
01-09-22	0.3	0.1
02-08-01	0.2	0.2
00-09-27	0.2	0.2
01-04-20	0.2	0.2
98-09-23	0.1	0.1

turn rate of 4.3 years for the whole of Barcelona, while the return rate was 16.3 years for one of the rain gauges.

In Figure 3.7, the rain gauge readings for two of these episodes are shown. The rainstorm presented in Figure 3.7 (a) caused severe flooding in the area of the city under study.

3.4 Summary

This chapter described modelling principles relating to sewer networks. The basic system model was obtained from the assumption that the network is composed of virtual tanks that correspond to the sewage storage capacity of a set of sewers in a given catchment. From that moment on, the network was (and will continue to be) considered a collection of tanks, connective links (main sewers) and gates (among other elements), which constitute a representation of the functional elements that constitute any sewer network. However, the set of elements for an entire system has an associated mathematical model that includes nonlinear expressions relating to nonlinear system behaviours – a fact that adds more modelling considerations which must be addressed when a constrained predictive control law is designed. Once the system model had been determined, a parameter calibration method using real data was described.

Finally, this chapter presented a brief description of a particular sewage system: Barcelona's sewer network. Its main problems were outlined and a case study was proposed, which is based on a representative portion of

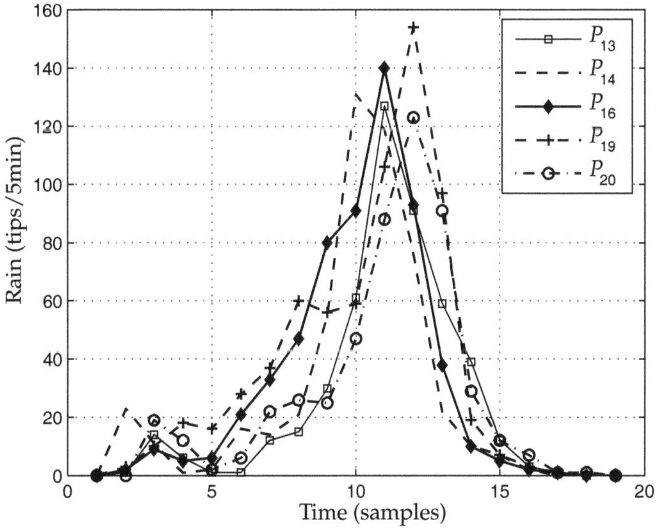

(a)

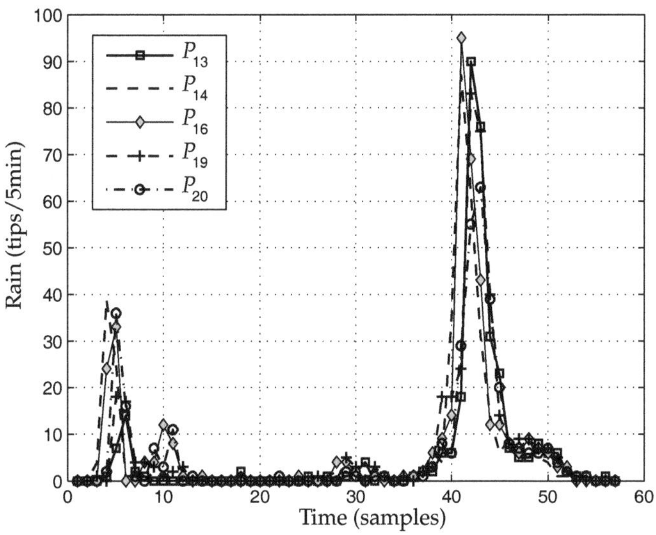

(b)

Figure 3.7 Examples of rain episodes occurred in Barcelona: (a) September 14, 1999, and (b) October 17, 1999. Each curve depicts data from a particular rain gauge P_i

the entire network. The BTC model is calibrated using real measurements from the network sensors (limnimeters) and data from rain episodes that occurred in the city of Barcelona between 1998 and 2002. The rain episodes used to simulate this system and to design the control strategies in following chapters were also presented.

An exhaustive mathematical description of the BTC as well as a description of a MATLAB® Simulink® tool for simulating the behaviour of the case study and for preliminary optimal control design can be found in Ocampo-Martinez (2004).

Part II
Model Predictive Control of Sewer Networks

Chapter 4
Formulating the Model Predictive Control Problem

4.1 General Considerations

One of the most effective and accepted control strategies for the sewage control problem is MPC. One early work which suggested that this approach could be applied to such a problem was Gelormino and Ricker (1994). In that work, an implementation of linear model predictive control of the Seattle urban drainage system was presented. Their results confirmed the effectiveness of the global predictive control law relative to conventional local automatic controls and heuristic rules that were used to control and coordinate the overall system. Other articles where predictive control ideas for wastewater systems were developed further include Marinaki and Papageorgiou (2001), Duchesne *et al.* (2004), Wahlin (2004), Caraman *et al.* (2007) and Cen and Xi (2009), among many others.

The predictive controller is usually thought to occupy the middle level of a hierarchical control structure. The states are estimated and the rain is forecast over the prediction horizon at the top of such a structure. This information is the input for the MPC problem. The outputs of the MPC controller are reference values for decentralised local controllers that implement the calculated set-points. See Marinaki and Papageorgiou (1997), Schütze *et al.* (2004), Marinaki and Papageorgiou (2005) and Brdys *et al.* (2008) for references where this hierarchical structure is followed.

As there are many control objectives associated with the sewer network control problem, the optimisation problem associated with the MPC controller also has multiple objectives. The most common approach used to solve multi-objective optimisation problems is to form a scalar cost function that consists of a linearly weighted sum of cost functions associated with each objective. When objectives have a priority (that is, when certain objectives are considered more important than others), then the aim is to reflect this importance through the appropriate selection of weights. However, finding appropriate weights is not a trivial problem, especially for large-scale control problems with multiple objectives, as in the case of sewer networks. Because cost functions have different numerical values in different scenarios, weights that are appropriate for one situation might not be appropriate for another. The weights therefore serve to normalise the cost functions as well as to organise their priority. Furthermore, in the case of sewer network control, some objectives are only relevant under specific circumstances. For instance, when there is little precipitation there is no risk of flooding, although stopping the release of untreated sewage maintains its importance as a control objective. Thus, focusing on the issue of flooding while selecting weights might not be an appropriate approach when this phenomenon is not present.

Generally, the selection of weights is done by heuristic trial and error procedures involving a lot of numerical simulations (see, *e.g.*, Tyler and Morari, 1999). This complicates and increases the cost of implementing predictive controllers for sewer networks. Furthermore, it becomes complicated to maintain the controllers and adapt to changes in the system, as the weights need to be revised in these cases.

An alternative to weight-based methods, the lexicographic approach, is based on assigning *a priori* different priorities to the different objectives and then optimising the objectives in order of priority. Establishing the priorities of the objectives using lexicographic minimisation is conceptually very simple, especially in sewer network control, and requires marginal implementation effort compared to the weight-based approach. This chapter discusses the application of lexicographic minimisation to eliminate the process of weight selection when designing MPC controllers for sewer networks. Lexicographic minimisation has been mentioned before in the context of MPC (see, *e.g.*, Tyler and Morari, 1999; Vada *et al.*, 2001; Kerrigan and Maciejowski, 2002; Aggelogiannaki and Sarimveis, 2006; Zheng *et al.*, 2007; Gambier, 2008; Padhiyar and Bhartiya, 2009), but few real applications – especially to large-scale systems – have been presented.

Keep in mind that only linear models of the sewer network are considered. The main reason for this is to keep the optimisation problem convex. Convexity is an important property that guarantees applicability of the MPC methodology to large-scale problems. The use of nonlinear models for the predictive control of urban drainage systems has also been reported (Marinaki and Papageorgiou, 1998). However, improvements in prediction achieved using nonlinear models should be compared to the uncertainty present due to the error involved in predicting the rain over the horizon. If the improvement due to the use of nonlinear models is marginal compared to the uncertainty in rain prediction, nonlinear models are difficult to justify, as the related predictive control optimisation problem is often non-convex, which leads to difficulties with convergence to local minima and numerical efficiency when large-scale problems are considered. In Previdi *et al.* (1999), higher order linear models for sewer systems are identified, with good results.

It should be pointed out that short-term rain prediction or *nowcasting* is an active field of research (Smith and Austin, 2000). Combining radar data, rain gauge measurements and advanced data processing has resulted in vastly improved rain prediction in recent times, and the potential use of this approach in the predictive control of urban drainage systems was pointed out in Yuan *et al.* (1999). However, it is beyond the scope of the current chapter to explore the trade-offs between the use of linear and nonlinear models in the context of modern rain prediction methods.

4.2 Control Problem Formulation

As the model and constraints are linear, the MPC controller presented in this chapter is designed using textbook formalisms (see, *e.g.*, Maciejowski, 2002; Goodwin *et al.*, 2005; Rawlings and Mayne, 2009). From the discussion presented in Chapter 3, the model of a sewer network can be written as

$$x_{k+1} = Ax_k + Bu_k + B_p d_k, \tag{4.1}$$

where x_k is the state vector that collects together the tank volumes v_i (both virtual and real), u_k represents the vector of manipulated flows $q_{u_i k}$, vector d_k corresponds to rain perturbations, and the constant matrices A, B and B_p are system matrices of suitable dimensions. Equation 4.1 is created using (3.2), (3.3), (3.5) and the topology of the sewer network. When the lower bounds of $q_{u_i k}$ are zero, the model constraints can be written as

$$Ex_k + Hu_k \leq b, \tag{4.2}$$

where E, H and b are matrices of suitable dimensions created using (3.6) and (3.4) as well as the range limitations of the manipulated flows $q_{u_i k}$.

The model presented in (3.2) is a first-order model relating inflows and outflows with a tank volume. In Previdi *et al.* (1999), higher order linear models were identified as a function of inflows and outflows. Good results were obtained, even when output error models were used for simulation. The control methodology presented there can be applied virtually unchanged if a more general linear filter is considered, for example one obtained from on-line calibration procedures, instead of the model in (3.2). In the software implementation, the states are expressed as affine functions of the changes in the control signal; *i.e.*,

$$\Delta u_k = u_k - u_{k-1}, \tag{4.3}$$

for a prediction horizon H_p. On the other hand, the control signal is only allowed to change over the control horizon H_u.

4.2.1 *Control Objectives*

The sewage system control problem has multiple objectives with varying priorities; (see Marinaki and Papageorgiou, 2005). There are many types of objectives, depending on the system design. In general, the most common objectives are related to the manipulation of the sewage in order to avoid undesirable sewage flows outside of the main sewers. Another type of control objective is related, for instance, to the control energy (*i.e.*, the energy cost of the movements of regulation gates). According to the literature on sewer networks, the main control objectives for the case study of this book are as follows (listed in order of decreasing priority):

- *Objective 1:* to minimise flooding in streets (virtual tank overflow) (f_1)
- *Objective 2:* to minimise flooding in links between virtual tanks (f_2)
- *Objective 3:* to maximise sewage treatment (f_3)
- *Objective 4:* to minimise control action (f_4).

A secondary purpose of the third objective is to reduce the volume in the tanks, in anticipation of future rainstorms. This objective indirectly reduces pollution to the receiving environment: if the WWTPs are used optimally according to the storage capacity of the network, then the pollution should be strongly minimised. Moreover, this objective can be complemented by applying a minimum volume condition to the real tanks at the end of the prediction horizon. This could be seen as a fifth objective.

4.2.2 *Cost Function Formulation*

Considering the control objectives mentioned before, it is necessary to define the criteria for minimising the performance indices. In this chapter, the 1-norm and ∞-norm are considered when the cost function (2.5a) of the OOP (2.5) is stated.

State Feedback MPC Statement in the 1-Norm and ∞-Norm Cases

Consider an alternative form of the OOP (2.5), expressed as

$$\min_{\mathbf{u}_k \in \mathscr{U}_{H_p}} J(\mathbf{u}_k, x_k, H_p) = \sum_{i=0}^{H_p} \|Qx_i\|_p + \sum_{i=0}^{H_p-1} \|Ru_i\|_p, \tag{4.4a}$$

subject to

$$x_{k+1} = Ax_k + Bu_k, \quad k = 0, \ldots, H_p - 1, \tag{4.4b}$$

$$x_k \in \mathbb{X} \subseteq \mathbb{R}^n, \tag{4.4c}$$

$$u_k \in \mathbb{U} \subseteq \mathbb{R}^m, \tag{4.4d}$$

where Q and R are weight matrices of suitable dimensions and $p - 1, \infty$. According to Campo and Morari (1986), (4.4) can be reformulated as the linear programming problem

$$\min_{\eta} \varepsilon_0^x + \ldots + \varepsilon_{H_p}^x + \varepsilon_0^u + \ldots + \varepsilon_{H_p-1}^u, \tag{4.5a}$$

subject to

$$-\mathbb{1}_n \, \varepsilon_k^x \leq \pm Q \left[A^k x_0 + \sum_{j=0}^{k} A^j B u_{k-1-j} \right], \tag{4.5b}$$

$$-\mathbb{1}_m \, \varepsilon_{k-1}^u \leq \pm R u_{k-1}, \quad k = 0, \ldots, H_p, \tag{4.5c}$$

where

$$\eta \triangleq \{\varepsilon_0^x, \ldots, \varepsilon_{H_p}^x, \varepsilon_0^u, \ldots, \varepsilon_{H_p-1}^u, u_0, \ldots, u_{Hp-1}\} \in \mathbb{R}^s,$$

with $s \triangleq (m+1)H_p + H_p + 1$, and where $\mathbb{1}$ corresponds to a vector of ones of suitable dimensions. Constraints (4.5b) and (4.5c) are componentwise, and $\pm$ implies one constraint for each sign.

Particular Formulation

Variables related to the first and second control objectives are overflow variables, which depend on the state. These variables can be treated as *slack variables* at the overflow constraints (see Gelormino and Ricker, 1994, for a similar approach). In the case of virtual tank overflow, these variables are expressed as

$$(\hat{v}_{ik+j|k} - \overline{v}_i)/S_i \leq \varepsilon_{i\ k+j|k}^v, \tag{4.6a}$$

$$0 \leq \varepsilon_{i\ k+j}^v \tag{4.6b}$$

for all tanks $i = 1 \ldots n$ and for $j = 1 \ldots H_p$. $\hat{v}_{ik+j|k}$ denotes the prediction of the state at time $k \in \mathbb{Z}_+$, j samples into the future. For the first two objectives, the vectors of slack variables are defined as

$$\Psi_v = [\varepsilon_{k+1|k}^v, \ \cdots \ , \varepsilon_{k+H_p|k}^v], \tag{4.7}$$

$$\Psi_{q_s} = [\varepsilon_{k|k}^{q_s}, \ \cdots \ , \varepsilon_{k+H_p-1|k}^{q_s}]. \tag{4.8}$$

Vector Ψ_{q_s} has $N_{q_s} H_p \times 1$ elements, where N_{q_s} is the number of overflow links. Notice that the slack variable $\varepsilon^v_{k|k}$ is not defined, as it depends on $\hat{v}_{k|k}$ or the measured state at time k, which cannot be affected by control actions. For the same reason, Ψ_{q_s} does not include variables for time $k + H_p$. The third and fourth objectives are expressed using the vectors

$$\Psi_{\mathrm{TP}} = [\overline{q}^{\mathrm{TP}} - q^{\mathrm{TP}}_{k|k}, \; \dots \; , \overline{q}^{\mathrm{TP}} - q^{\mathrm{TP}}_{k+H_p-1|k}], \tag{4.9}$$

$$\Psi_{\Delta u} = [\Delta u_{k|k}, \; \dots \; , \Delta u_{k+H_p-1|k}]. \tag{4.10}$$

The variable $q^{\mathrm{TP}}_{k|k+i}$ is a vector containing the flows to the treatment plants located in the network, $\overline{q}^{\mathrm{TP}}$ is its maximum, and Δu is a vector containing the changes in control action between samples, which is defined within this framework as $\Delta u = q_{u_i k} - q_{u_i k-1}$. Using variables (4.7)–(4.8) and (4.9)–(4.10), the control objectives described above can be formulated mathematically as the minimisation of the following cost functions:

$$f_1 = \|\Psi_v\|_\infty, \quad f_2 = \|\Psi_{q_s}\|_\infty, \quad f_3 = \|\Psi_{TP}\|_1 \quad \text{and} \quad f_4 = \|\Psi_{\Delta u}\|_1. \tag{4.11}$$

The ∞-norm is used because it is desirable to minimise the maximum flooding over the prediction horizon. For wastewater treatment, on the other hand, the total treated volume of sewage is more important than peaks in its behaviour.

4.2.3 *Control Problem Constraints*

The physical constraints of the system presented in (3.6) and (3.4) for the case of real tanks are added as constraints in the optimisation problem associated with MPC controller design. For each slack variable ε, constraints in (4.6) are also included.

4.3 Multi-objective Optimisation

The optimisation problem associated with MPC controller design is multi-objective. Recent surveys of multi-objective optimisation can be found in Marker and Arora (2004), Chinchuluun and Pardalos (2007) and Branke *et al.* (2008). In general, such a problem can be formulated in the following way:

$$\min_{z \in \mathscr{Z}} \quad [f_1(z), f_2(z), \cdots, f_r(z)], \tag{4.12}$$

where $z \in \mathscr{Z}$ is a vector containing the optimisation variables, $\mathscr{Z} \subseteq \mathbb{R}^p$ is the admissible set of optimisation variables, and f_i are scalar-valued functions of z.

Definition 4.1 (Pareto-optimal Solution). A solution z^* is said to be a *Pareto-optimal* if and only if another $z \in \mathscr{Z}$ does not exist such that $f_i(z) \leq f_i(z^*)$ for all $i = 1, \cdots, r$ and $f_j(z) < f_j(z^*)$ for at least one index j. In other words, a solution is Pareto-optimal if an objective f_i can be reduced at the expense of increasing at least one of the other objectives. In general, there may be many Pareto-optimal solutions to an optimisation problem.

A common approach to solving multi-objective optimisation problems is scalarisation (Chinchuluun and Pardalos, 2007). This means converting the problem into a single-objective optimisation problem with a scalar-valued objective function. A common way to obtain a scalar objective function is to form a linearly weighted sum of r functions f_i as

$$\sum_{i=1}^{r} w_i f_i(z). \tag{4.13}$$

The priorities of the objectives are reflected in the weights w_i. Although this type of scalarisation is widely used, it has serious drawbacks (Miettinen, 1999). Practical drawbacks to this approach, specifically for large-scale systems, are detailed in Benali *et al.* (2007).

If the objectives are prioritised, there is a unique solution on the Pareto surface where this order is respected (see Kerrigan and Maciejowski, 2002, and references therein). Let the objective functions be arranged in order of priority from the most important f_1 to the least important f_r.

Definition 4.2 (Lexicographic Minimiser). A given $z^* \in \mathscr{Z}$ is a lexicographic minimiser of (4.12) if and only if there is not a $z \in \mathscr{Z}$ and an i^* that satisfies $f_{i^*}(z) < f_{i^*}(z^*)$ and $f_i(z) = f_i(z^*), i = 1, \ldots, i^* - 1$. The corresponding solution $f(z^*)$ is the lexicographic minimum.

One interpretation of Definition 4.2 is that a solution is a lexicographic minimum if and only if an objective f_i can be reduced only at the expense of increasing at least one of the objectives of greater priority $\{f_1, ..., f_{(i-1)}\}$. Hence, a lexicographic solution is a special type of Pareto-optimal solution

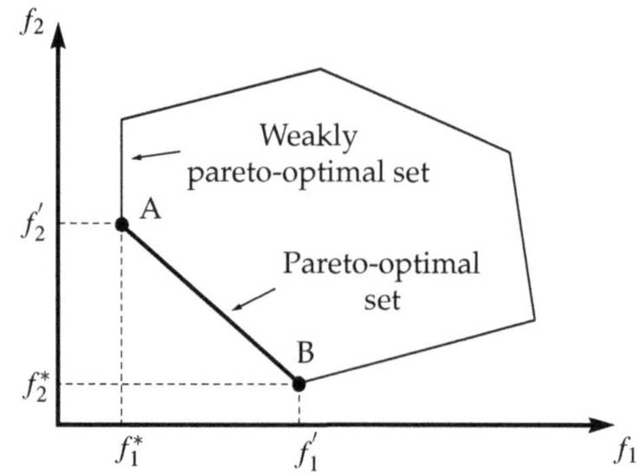

Figure 4.1 Conceptual scheme of the Pareto-optimal set

that takes into account the order of the objectives. This hierarchy defines an order for the objective function which establishes that a more important objective is infinitely more important than a less important objective.

In order to illustrate the concepts collected in Definitions 4.1 and 4.2, Figure 4.1 depicts the Pareto-optimal set for a problem with two performance objectives f_1 and f_2, where f_1 is more important than f_2. According to the prioritisation given to those objectives, and employing the lexicographic approach, objective f_1 is minimised until its optimal value f_1^*. Note that once $f_1 = f_1^*$, objective f_2 can take all of the values of the *weakly Pareto-optimal set*. Therefore, following the pre-established lexicographic order, f_2 can be minimised until the value f_2' (point A in the figure). Hence, point A corresponds to the optimal value for the considered optimisation problem. By contrast, if the control objectives have the opposite prioritisation, the point (f_1', f_2^*), denoted in Figure 4.1 as B, would be the optimal value of the optimisation problem. Also notice that, in the case of a prioritisation based on weights, all of the solutions of this problem (which depend on the combination of the weights) will lie in the Pareto-optimal set described by the line between A and B.

A standard method for finding a lexicographic solution is to solve an ordered sequence of single-objective constrained optimisation problems. After ordering, the most important objective function is minimised subject to the original set of constraints. If this problem has a unique solution, it is the solution of the whole multi-objective optimisation problem. Otherwise, the second most important objective function is minimised. Now, in addition to the original constraints, a new constraint is added. This new constraint is there to guarantee that the most important objective function preserves its optimal value. If this problem has a unique solution, it is the solution of the original problem. Otherwise, the process continues as above. Algorithm 4.1 formally defines this sequential solution method of finding the lexicographic minimum of (4.12).

Algorithm 4.1 Lexicographic multi-objective optimisation using the sequential solution method

1: $f_1^* = \min_{z \in \mathscr{Z}} f_1(z)$
2: **for** $i = 2$ to r **do**
3: $\quad f_i^* = \min \left\{ f_i(z) | f_j(z) \leq f_j^*, j = 1, ..., i-1 \right\}$
4: **end for**
5: Determine the lexicographic minimiser set as: $z^* \in \left\{ z \in \mathscr{Z} | f_j(z) \leq f_j^* , j = 1, \cdots, r \right\}$

Other approaches for finding the lexicographic minimum aside from the sequential solution approach have also been presented. Tyler and Morari (1999) and Kerrigan *et al.* (2000) showed how the sequential solution approach could be replaced by a single MIP problem. Vada *et al.* (2001) described how the weights for scalar objective function (4.13) could be found such that the solution of the scalar problem would be a lexicographic minimum. The weights are found by solving a multi-parametric LP (mpLP) problem. The parameters are the components of the measured state $\hat{x}_{k|k}$ of the system to be controlled. In the current case, where large-scale systems are considered and where the disturbances (rain) are predicted over the prediction horizon and included in the optimisation problem, the number of parameters for which the mpLP would be solved off-line would be related to not only the number of states but also the number of disturbances multiplied by the length of the prediction horizon. This would lead to a very large multi-parametric programming problem and the advantages over using the sequential solution approach would be lost.

The sampling time in sewer network control is generally large (on the order of several minutes). This fact gives plenty of time for modern LP solvers to solve many large-scale problems, enabling the use of the sequential solution approach to lexicographic minimisation to obtain the control signal.

In order to perform a thorough comparison of the lexicographic minimisation approach with the weight-based approach to prioritising the control objectives, the performance of the closed-loop system was compared for 15 rain episodes of different intensities that are representative of Barcelona weather. The strategies were only compared in a simulation, as it is impossible to repeat real experiments, for obvious reasons.

4.4 Closed-loop System Configuration

4.4.1 Model Definition

Notice that the approach discussed in the current chapter considers a linear model of the system. Even if the tanks (real and virtual) are modelled by first-order linear models, the weirs R_i in Figure 3.4 can not be modelled adequately with a linear expression. Hence, the assumption made in this chapter is that these elements can be considered to be redirection gates, which makes it possible to show the proposed methodology. Figure 4.2 depicts the reconfigured system, and the modified elements are termed *manipulated overflow elements*. Keep in mind that all particular descriptions, concepts and parameter values for the BTC defined in Chapter 3 remain the same for this modified case study.

4.4.2 Simulation of Scenarios

For simplicity, the second objective was omitted in the simulation of the case study for this chapter. This is advantageous to the weighted approach, as extra control objectives mean just one more optimisation in the lexicographic case, while they make the selection of weights more difficult in the weight-based approach.

MPC controller tuning based on lexicographic minimisers was designed using Algorithm 4.1 with the cost functions given in (4.11). Notice that the f_i have different norms, namely the ∞-norm and the 1-norm. Both of these norms result in a linear program to solve the MPC problem, which in turn means that when the result of an optimisation is passed along to the subsequent optimisation in Algorithm 4.1 in order to generate a new constraint, linear constraints can be used.

The weight-based approach used the same cost functions f_i to express the control objectives, but the control signals of the MPC controller were found as the solution when cost function (4.13) was minimised. For an exhaustive comparison, a range of ratios between the first two weights, w_1 and w_3, were considered. At one extreme of this range ($w_1/w_3 = 200$), the first objective became the same value as that obtained when the other terms of the cost functions were not present. At the other extreme of this range ($w_1/w_3 = 0.4$), the first objective started to suffer and the performance degraded, which even caused a reversal of priorities between objectives. Notice that the numerical values of f_i differ significantly among the scenarios. The values of the other weights were carefully selected to ensure that the numerical values of the terms $w_i f_i$ would be much smaller in the scenarios considered.

In order to compare strategies for a rain episode, the best performance from the range of w_1/w_3 ratios was compared to the performance of the lexicographic approach. The values shown in Table 4.1 correspond to this selection. Thus, no one ratio was considered for all scenarios; the optimal weight was selected after simulating each scenario.

The control strategies/tunings were compared by simulating the closed-loop system for each rain episode. The model used for simulation (open-loop model) was the same as that used for the model predictive controller.

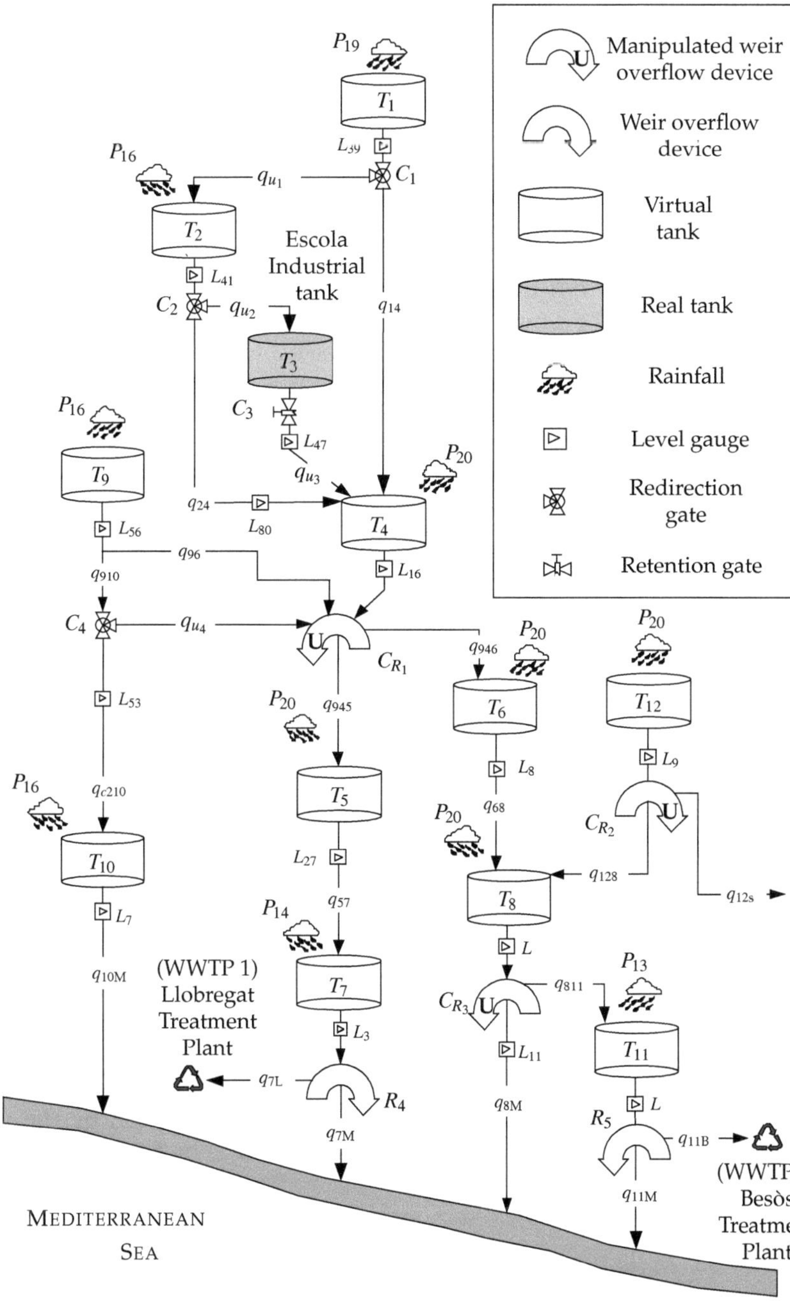

Figure 4.2 BTC where the principal weirs are considered manipulated overflow elements

The duration of each simulation was chosen to be 80 samples or around 6.5 h, as the rainstorm duration generally peaked at around ten samples or 50 min. The tanks were empty at the beginning of each scenario. The rainstorm peaks generally occurred between samples 1 and 25. Some of the rain episodes considered could have longer periods of significant rain.

The prediction and control horizons were selected to be six samples or 30 min, which corresponds to the *time of concentration*[1] for the Barcelona sewer network. This was chosen based on the heuristic knowledge of the engineers at CLABSA and field tests performed in the sewer network. Another reason for selecting these prediction and control horizon values is that the predictions provided by the sewer network model used become less reliable for larger time horizons. In this context, Figures 4.3 (a) and 4.3 (b) show comparisons between real sewage levels (from real data) and predicted sewage levels (obtained using the model described in Section 3.1) for the output sewer mains of virtual tanks T_1 and T_2, respectively, when the model was used to predict six steps ahead. It is apparent that the fits obtained with the proposed modelling approach are not as accurate as those in Figure 3.6, where the predictions were made for one step ahead. Moreover, it can also be seen that whether constant rain or known rain is considered during the prediction also affects the quality of the prediction.

4.4.3 Criteria for Comparison

To compare the strategies for the simulation scenarios, values related to the control objectives were calculated for each scenario. For the first objective, the maximum flooding across the whole scenario (that is, the maximum value of $\|\Psi_v\|_\infty$ for the whole scenario) was compared for each control strategy. For the second objective, the total volume of treated water was added up for the scenario. The water released to the receiving environment was added up in the same way. These values, obtained for each control strategy, were compared for each rain episode. The results are summarised in Table 4.1 (where TS denotes *treated sewage*). Finally, to determine which control strategy was the best, the values related to the control objectives were compared in a lexicographic manner; *i.e.*, in the order given by the pre-established priorities. If the first values were found to be equal, the second values were compared, and so on.

[1] The time of concentration of a sewer network is the time required for water to travel from the most remote catchment to its outlet to the environment (Mays, 2004).

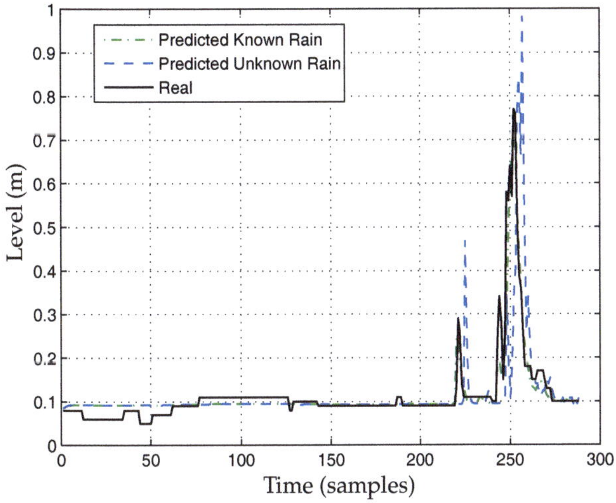

(a)

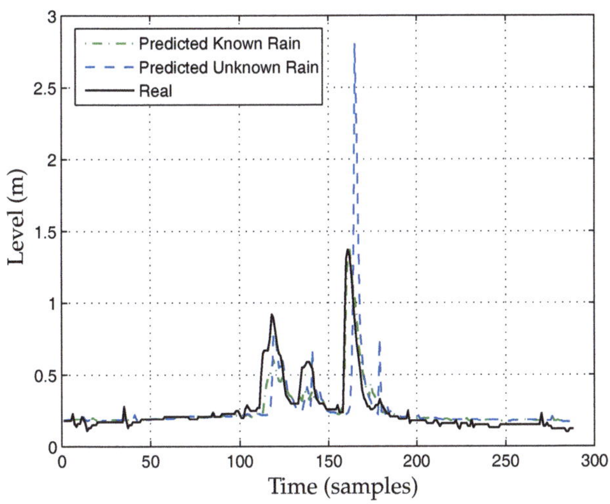

(b)

Figure 4.3 Results of model calibration using the approach given in Section 3.2 for predicting six steps ahead: (a) output level at T_1, and (b) output level at T_2

4.5 Discussion of the Results

The main results obtained, as derived from the numerical results summarised in Table 4.1, are now discussed. Significant performance improvements were obtained in the second objective when lexicographic minimisa-

Table 4.1 Performance results obtained with the control objectives considered

Rain episodes	Weighted approach Flooding (cm)	Pollution (hm^3)	TS (hm^3)	Lexicographic programming Flooding (cm)	Pollution (hm^3)	TS (hm^3)
99-09-14	11.5	200	292	11.5	197	295 (1.0%)
02-07-31	6	236	310	6	226	312 (0.6%)
02-10-09	3.6	384	504	3.2	380	508 (0.8%)
99-09-03	0	52	222	0	48	225 (1.3%)
99-10-17	0.3	71	274	1	70	275 (0.4%)
00-09-28	1.2	108	271	1.4	109	271 (0%)
98-10-05	0	3	85	0	0	89 (4.5%)
98-09-25	0	7	299	0	4	304 (1.6%)
98-10-18	0	5	125	0	0	130 (3.8%)
00-09-19	0	3	64	0	0	67 (4.5%)
01-09-22	0	30	181	0	28	183 (1.1%)
02-08-01	0	7	259	0	0.4	266 (2.6%)
00-09-27	0	8	101	0	0	109 (7.3%)
01-04-20	0	42	224	0	39	228 (1.7%)
98-09-23	0	2	70	0	0	72 (2.7%)

tion was used. No performance improvements with regard to the first objective were observed. Maximum flooding remained the same for the two control strategies, with the exception of two cases (episodes 99-10-17 and 00-09-28). The average percentage increase in treated sewage was 2.3%. For rain episodes where the return rate was lower than 0.3, the increase was 3.4%. The percentage increases are shown in brackets in the last column of Table 4.1.

The reason for the increase in sewage treatment was that the virtual tanks were used more efficiently to keep the levels in tanks 7 and 11 higher. This in turn enabled the inflow into the WWTPs to increase, as the outflow from these tanks could be redirected to those plants.

The improvement in sewage treatment led to a significant decrease in the pollution released into the environment. It can be seen in Table 4.1 that, for six rain episodes, the sewage released into the environment was reduced to zero when lexicographic minimisation was used. The pollution released (flow and total volume) into the receiving environment is shown in Figure 4.4 for episode 00-09-27. No performance improvement was observed for the first objective because the same performance could always be achieved by selecting a sufficiently large weight ratio w_1/w_3.

In episode 99-10-17, the maximum flooding in the virtual tanks across the whole scenario was higher when lexicographic minimisation was used compared to when the weight-based approach was used. The reason for this is explained in Figure 4.5. The rain episode has two peaks: a smaller one at time 5 and a larger one at time 43. Looking at the levels in the tanks, the level is higher at time 40 when the second peak starts in the lexicographic case.

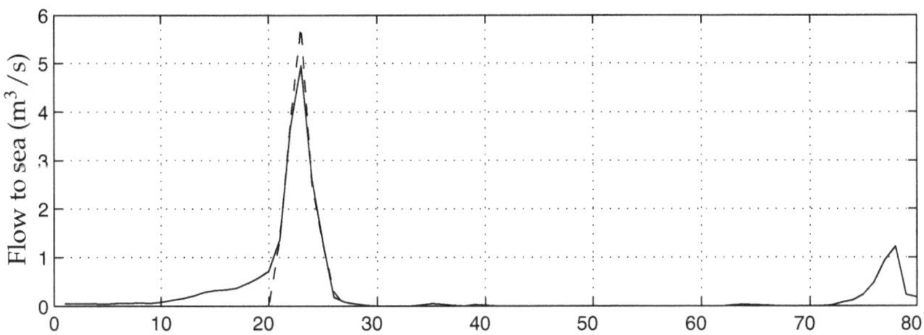

(a)

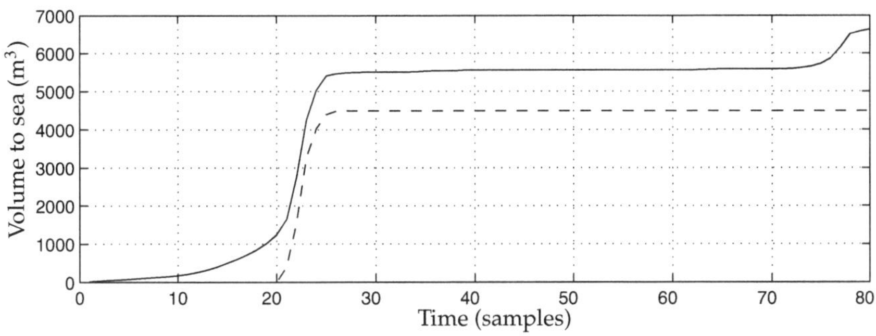

(b)

Figure 4.4 (a) Flow and (b) total volume released into the environment for the rain episode 00-09-27. *Solid curve,* weight-based prioritising; *dashed curve,* lexicographic prioritising

This causes the flooding to increase considerably around time 46. The other scenario where lexicographic minimisation performs less well (episode 00-09-28) also has double peaks.

The double-peak episodes are complicated, because the poor performance observed in these cases is actually related to the quality of rain prediction in the scenarios. The controller based on lexicographic minimisation is operating correctly until the second peak arrives. It accumulates sewage with the purpose of maximising sewage treatment. On the other hand, when the second peak arrives, this behaviour is counterproductive. Note that if the objective relating to sewage treatment is dropped from the objective functions, both of the controllers could reduce flooding to zero for episode 99-10-17, and reduce it substantially for the other episode.

The basic problem is that the first two objectives encourage opposite behaviours in the controllers. The optimal behaviour for the first objective is to have the virtual tanks as empty as possible so that they have the capacity

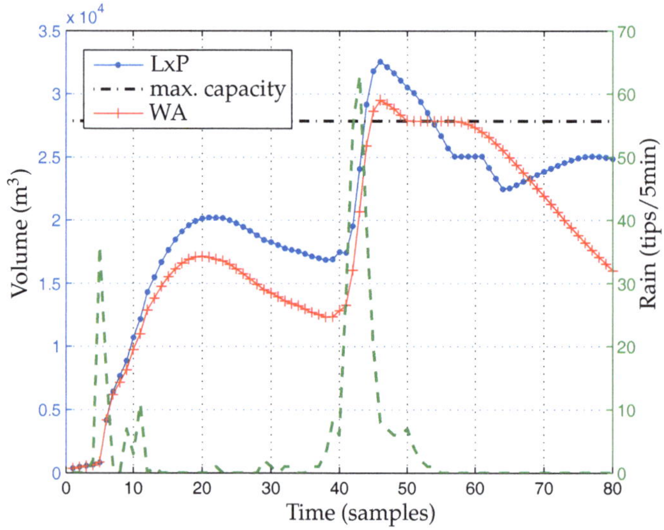

Figure 4.5 Case where lexicographic minimisation exhibits poor performance. The *dashed curve* is the rain intensity (*right axis*)

to be able to receive new surges of rainwater. The second objective strives to store sewage in the system to maximise sewage treatment. In the double-peak episodes, the controller based on the proposed lexicographic approach suffers due to its superior ability to achieve the second objective in the absence of flooding. If prediction of rain could be improved to the point of being able to recognise multi-peak episodes, a remedy to the problem described would be to simply drop the second objective until it is sure that flooding danger is not present.

Initially, two cases were considered for predicting rain $d(k)$ over the prediction horizon (30 min). In the first case, $d(k)$ was assumed to be equal to the last measurement over the whole horizon. In the other case, the rain was assumed to be accurately predicted over the horizon. Real rain predictions for an urban area of the type considered here would lie somewhere between these two cases: they would be better than assuming the rain to be constant over the horizon, but worse than accurate prediction. The MPC strategy allows the control actions to be updated to account for the real state of the system and the precipitation intensity at each sampling time. This state feedback reduces the effect of the rain prediction error on the control performance. In fact, the difference between the two cases was found to be small.

4.6 Summary

This chapter presented the lexicographic approach as a technique for achieving multi-objective optimisation, which arises during the application of MPC to sewer networks. The possible benefits of using lexicographic minimisation were demonstrated using a modified version of the BTC, where the most important rain episodes that occurred over an interval of four years were studied and the control performance was compared with that of the traditional weight-based approach to expressing the priorities of control objectives. It was shown that the performances of the two approaches were similar in relation to the objective with the highest priority, while performance of the lexicographic minimisation approach was better for objectives of lower priority. For light rain episodes, sewage treatment could be improved by 3.5%. The average improvement for all scenarios was 2.4%.

The numerical values of these performance improvements should be viewed in the context of the high infrastructural costs of urban wastewater systems. MPC based on lexicographic minimisation is conceptually very simple and requires marginal implementation effort compared to the weight-based approach once the system model is available. As the sewer networks of big cities continue to increase in complexity, due to their use of multiple reservoirs and actuators (gates), there will be a substantial increase in the number of criteria that must be accounted for in the global optimisation function, and so a great deal of off-line effort will be needed to find a good compromise when tuning all of the parameters in the weight-based approach. The lexicographic approach is therefore a very interesting tool for efficiently solving these kinds of problems. Its only disadvantage is the increased number of optimisation problems that are required to obtain the lexicographic minimum. However, if a linear model is used, modern convex optimisation routines can easily handle the large-scale systems that occur in the predictive control of urban drainage networks.

Note that the assumption of a linear system model limits the accuracy of the description of the real behaviour of the sewer network. This fact motivates to find different model representations that cover all of the dynamics without losing the advantages of implementing MPC in linear systems. Subsequent chapters deal with this problem and propose different ways of modelling a sewer network in order to design optimisation-based controllers.

Chapter 5
Predictive Control Problem Formulation and Hybrid Systems

Chapter 4 introduced the application of predictive control to sewer networks, showed how this resulted in significant improvements in the performance of the closed-loop system, and described the advantages of this control strategy for this type of networked system. However, the linear nature of the network model did not accurately reflect some of the dynamics and behaviours of some of its components (*e.g.*, weirs, tank overflows).

Therefore, a modelling methodology that can reflect the real dynamics of those components without abandoning the linear framework and losing all of the advantages of MPC for linear systems is needed. This fact motivated the proposal of a modelling methodology where nonlinear dynamics of the form seen in (3.9) can be taken into account for some of the sewer network elements considered. These dynamics were found to be mode commutations, where a logic variable determines the continuous behaviour of a particular set of elements, and hence the new global behaviour of the whole sewer network. This mixture of continuous dynamics and logical events leads to hybrid systems.

This chapter deals with the modelling of a generic sewer network using the hybrid systems framework. The system is decomposed into functional subsystems in order to get a clear picture of the *level of hybridity* related to each component. Using the MLD form – the well-known framework for modelling and controlling systems described by interdependent physical laws, logic rules, and operating constraints (Bemporad and Morari, 1999) – the entire model is expressed via a discrete linear state-space representation. The resulting MLD form is used to design an MPC law, which is computed by solving a discrete optimisation problem. Both the modelling methodology and the control design are applied across the BTC, showing the advantages of the proposed controller design and the improvement in performance gained by using a closed-loop system rather than an open-loop (*i.e.*, with no control law) system.

5.1 Hybrid Modelling Methodology

The presence of intense precipitation causes some sewer mains and virtual tanks to surpass their limits. When this happens, any excess above the maximum volume flows to another tank downstream. In this way, temporary flow paths are triggered that depend on the system state and inputs. Since this behaviour is observed in most parts of the sewer network, a modelling methodology is needed that can consider and incorporate overflows and other logical dynamics. The approach presented in this chapter only considers the most common constitutive elements of a sewer network. Other elements such as pumping stations can be easily modelled and added to the model of the networked system using the proposed hybrid modelling methodology.

The system exhibits hybrid behaviour in the flow links between tanks, in the tanks themselves (either virtual or real tanks), in the redirection gates, and in the weirs. Network sensors (rain and level gauges) can also be represented as hybrid systems due to their internal dynamics, but they are not taken into account in this manner in this book. The network model is also divided into functional parts in order to make it easier to determine the logical variables and their relations to other system variables. These parts are *virtual tanks* (VT), *real tanks with input gates* (RTIG), *sewage pipes* (SP) and *redirection gates* (RG). In this section, each element will be described and its equations within the MLD framework expressed. In order to obtain the MLD form, the following equivalences are used:

$$[f(x) \leq 0] \longleftrightarrow [\delta_k = 1] \quad \text{is true iff} \quad \begin{cases} f(x) \leq M(1-\delta_k), \\ f(x) \geq \varepsilon + (m-\varepsilon)\delta_k, \end{cases} \tag{5.1}$$

and

$$z_k = \delta_k f(x) \quad \text{is true iff} \quad \begin{cases} z_k \leq M\delta_k, \\ z_k \geq m\delta_k, \\ z_k \leq f(x) - m(1-\delta_k), \\ z_k \geq f(x) - M(1-\delta_k), \end{cases} \tag{5.2}$$

where $M, m \in \mathbb{R}$ are the upper and lower bounds on the linear function $f(x)$ for $x \in \mathbb{X}$. $\varepsilon > 0$ is the numerical tolerance of the computer (see Bemporad and Morari, 1999; Torrisi and Bemporad, 2004).

5.1.1 Virtual Tanks (VT)

This element is modelled as a hybrid system, with the following behaviour accounted for. When the maximum volume in the virtual tank is reached,

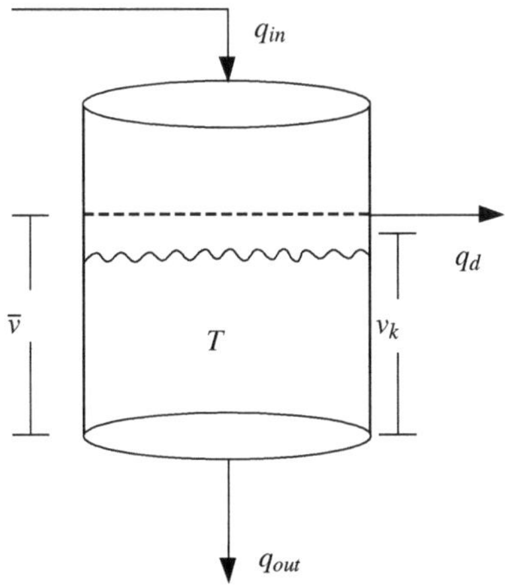

Figure 5.1 Scheme of a virtual tank

the excess above this maximum amount is redirected to another tank downstream. This phenomenon can be expressed mathematically as

$$q_{d\,k} = \begin{cases} \frac{(v_k - \overline{v})}{\Delta t} & \text{if} \quad v_k \geq \overline{v}, \\ 0 & \text{otherwise}, \end{cases} \tag{5.3}$$

$$q_{out\,k} = \begin{cases} \beta \overline{v} & \text{if} \quad v_k \geq \overline{v}, \\ \beta v_k & \text{otherwise}, \end{cases} \tag{5.4}$$

where v_k corresponds to the tank volume (system state) and $\overline{v}$ is its maximum volume capacity. The flow $q_{d\,k}$ corresponds to the tank overflow; see Figure 5.1. To get a feasible solution to the optimisation problem behind MPC controller design, the virtual tank volume must not be bounded by hard constraints.

Hence, in order to obtain the corresponding MLD form, the overflow condition is considered by defining the logical variable

$$[\delta_k = 1] \longleftrightarrow [v_k \geq \overline{v}], \tag{5.5}$$

which implies that the flows $q_{d\,k}$ and $q_{out\,k}$ are defined as

$$\begin{aligned} z_{1\,k} &= q_{d\,k} \\ &= \delta_k \left[\frac{(v_k - \overline{v})}{\Delta t} \right], \end{aligned} \tag{5.6}$$

$$\begin{aligned} z_{2\,k} &= q_{out\,k} \\ &= \delta_k \beta \overline{v} + (1 - \delta_k) \beta v_k. \end{aligned} \tag{5.7}$$

The corresponding difference equation for the tank as a function of the auxiliary variables is written as

$$v_{k+1} = v_k + \Delta t[q_{in_k} - z_{1k} - z_{2k}], \tag{5.8}$$

where q_{in_k} is the tank inflow, z_{1k} is related to the tank overflow, and z_{2k} is related to the tank output. Note that q_{in_k} collects all of the inflows to the tank, which could be outflows from tanks located downstream, link flows, overflows from other tanks and/or sewers and rain inflows.

Hence, the MLD expression (2.8) for this element, using the tank volume as the system output, can be written as follows:

$$v_{k+1} = v_k + [\Delta t]\, q_{in_k} + \begin{bmatrix} -\Delta t & -\Delta t \end{bmatrix} \begin{bmatrix} z_{1k} \\ z_{2k} \end{bmatrix}, \tag{5.9a}$$

$$y_k = v_k, \tag{5.9b}$$

$$\begin{bmatrix} -M_v + \overline{v} \\ \overline{v} \\ (M_v - \overline{v})/\Delta t \\ \overline{v}/\Delta t \\ -\overline{v}/\Delta t \\ -(M_v - \overline{v})/\Delta t \\ \beta\overline{v} \\ \beta(M_v - \overline{v}) \\ -\beta(M_v - \overline{v}) \\ \beta\overline{v} \\ 0 \\ 0 \end{bmatrix} \delta_k + \begin{bmatrix} 0 & 0 \\ 0 & 0 \\ -1 & 0 \\ 1 & 0 \\ -1 & 0 \\ 1 & 0 \\ 0 & -1 \\ 0 & 1 \\ 0 & -1 \\ 0 & 1 \\ 0 & 0 \\ 0 & 0 \end{bmatrix} \begin{bmatrix} z_{1k} \\ z_{2k} \end{bmatrix} \leq \begin{bmatrix} -1 \\ 1 \\ -1/\Delta t \\ 1/\Delta t \\ 0 \\ 0 \\ 0 \\ 0 \\ -\beta \\ \beta \\ 1 \\ -1 \end{bmatrix} v_k + \begin{bmatrix} \overline{v} \\ 0 \\ M_v/\Delta t \\ 0 \\ 0 \\ 0 \\ 0 \\ \beta M_v \\ 0 \\ 0 \\ 0 \\ M_v \end{bmatrix}, \tag{5.9c}$$

where M_v is related to the maximum value of the state variable, which would be unbounded (*i.e.*, $M_v \to \infty$) for virtual tanks.

5.1.2 *Real Tanks with Input Gates (RTIG)*

As mentioned previously, real tanks are elements that are designed to retain sewage when severe weather events occur. For this reason, both the tank inflow and outflow can be controlled. On the other hand, the tank inflow is constrained by the current volume stored in the tank, by its maximum volume capacity, and by the tank's outflow. Since these tanks are considered to have no overflow capabilities, the inflow is pre-manipulated using a redirection gate – hence the consideration of this component during the modelling of real tanks. Figure 5.2 depicts a scheme for this element.

Figure 5.2 Scheme of a real tank with an input gate

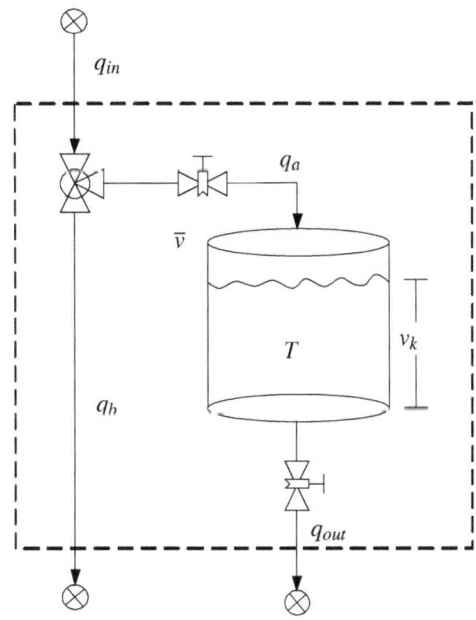

Element Model with Uncontrolled Flows

When an open loop[1] is modelled, the flow through the pipe q_a (where the pipe has a given maximum capacity) is defined by

$$q_{a1k} = \begin{cases} \overline{q}_a & \text{if} \quad q_{in k} \geq \overline{q}_a, \\ q_{in k} & \text{otherwise}, \end{cases} \tag{5.10}$$

where $q_{in k}$ is the inflow to the element and $\overline{q}_a$ represents the maximum flow through sewer q_a. However, if the inflow q_{a1k} causes the real tank to overflow, this inflow must be reduced to a flow equivalent to the volume needed to fill the tank. That is,

$$q_{a2k} = \begin{cases} v_{rk} & \text{if} \quad q_{a1k} \geq v_{rk}, \\ q_{a1k} & \text{otherwise}, \end{cases} \tag{5.11}$$

where $v_{rk} = \left(\frac{(\overline{v}-v_k)}{\Delta t} + q_{out k}\right)$ and $q_{out k}$ denotes the tank outflow. The MLD form for the given initial assumptions is obtained as follows. Defining

$$[\delta_{1k} = 1] \longleftrightarrow [q_{in k} \geq \overline{q}_a], \tag{5.12}$$

it is possible to obtain an auxiliary continuous variable

[1] This case represents the determination of a model for an autonomous system; *i.e.*, with no manipulated variables. Here, sewage flows through the mains due to gravity.

$$\begin{aligned} z_{1k} &= q_{ak}, \\ &= \delta_{1k}\overline{q}_a + (1-\delta_{1k})q_{in\,k}. \end{aligned} \tag{5.13}$$

For the auxiliary flow q_{a2k}, the following logical variable is defined:

$$[\delta_{2k} = 1] \longleftrightarrow \left[z_{1k} - q_{out\,k} \geq \frac{(\overline{v} - v_k)}{\Delta t} \right], \tag{5.14}$$

and hence the auxiliary variable equivalent to the new real tank inflow is expressed as

$$\begin{aligned} z_{2k} &= q_{a2k}, \\ &= \delta_{2k}\left[\frac{(\overline{v} - v_k)}{\Delta t} + q_{out\,k} \right] + (1-\delta_{2k})z_{1k}. \end{aligned} \tag{5.15}$$

Finally, due to mass conservation at the included input gate, q_{bk} is defined as $q_{bk} = q_{in\,k} - z_{2k}$, and the corresponding difference equation for the tank as a function of the auxiliary variables is

$$v_{k+1} = (1-\beta\Delta t)v_k + \Delta t z_{2k}. \tag{5.16}$$

Using the tank volume as the system output, MLD expression (2.8) is written as

$$v_{k+1} = [1-\beta\Delta t]\, v_k + \left[0\ \Delta t\right] \begin{bmatrix} z_{1k} \\ z_{2k} \end{bmatrix}, \tag{5.17a}$$

$$y_k = v_k, \tag{5.17b}$$

and inequality (2.8c), which actually collects together 12 linear inequalities:

$$E_1 = \left[1\ {-1}\ 0\ 0\ {-1}\ 1\ 0\ 0\ 0\ 0\ 0\ 0\right]^T, \tag{5.17c}$$

$$E_{21} = \begin{bmatrix} -\overline{q}_a & (\overline{q}_{in} - \overline{q}_a) & 0 & 0 & (\overline{q}_{in} - \overline{q}_a) & \overline{q}_a \\ 0 & 0 & -\frac{\overline{v}}{\Delta t} & (\overline{q}_{in} - \beta\overline{v}) & 0 & 0 \end{bmatrix}, \tag{5.17d}$$

$$E_{22} = \begin{bmatrix} -\overline{q}_a & -(\overline{q}_{in} - \overline{q}_a) & 0 & 0 & 0 & 0 \\ 0 & 0 & -\frac{\overline{v}}{\Delta t} & (\overline{q}_{in} - \beta\overline{v}) & 0 & 0 \end{bmatrix}, \tag{5.17e}$$

$$E_2 = \left[E_{21}\ E_{22}\right]^T, \tag{5.17f}$$

$$E_3 = \begin{bmatrix} 0 & 0 & -1 & 1 & -1 & 1 & -1 & 1 & 1 & -1 & 0 & 0 \\ 0 & 0 & 0 & 0 & 0 & 0 & 0 & 0 & -1 & 1 & -1 & 1 \end{bmatrix}^T, \tag{5.17g}$$

$$E_4 = \left[0\ 0\ \tfrac{1}{\Delta t} - \beta\ -\left(\tfrac{1}{\Delta t} - \beta\right)\ 0\ 0\ 0\ 0\ 0\ 0\ \tfrac{1}{\Delta t} - \beta\ -\left(\tfrac{1}{\Delta t} - \beta\right)\right]^T, \tag{5.17h}$$

$$E_{51} = \left[-\overline{q}_a\ \overline{q}_{in}\ -\tfrac{\overline{v}}{\Delta t}\ \left(\overline{v}\left(\tfrac{1}{\Delta t} - \beta\right) + \overline{q}_{in}\right)\ (\overline{q}_{in} - \overline{q}_a)\ \overline{q}_a\right], \tag{5.17i}$$

$$E_{52} = \left[-\overline{q}_a\ \overline{q}_a\ (\overline{q}_{in} - \beta\overline{v})\ \tfrac{\overline{v}}{\Delta t}\ -\tfrac{\overline{v}}{\Delta t}\ \tfrac{\overline{v}}{\Delta t}\right], \tag{5.17j}$$

$$E_5 = \left[E_{51}\ E_{51}\right]^T. \tag{5.17k}$$

Element Model with Controlled Flows

Depending on the use of this element in either the simulation or the prediction model, the tank inflow and its outflow can be manipulated. Hence, two possibilities are considered and outlined below:

1. The RTIG element is used to build a global hybrid model for simulation within a control loop. In this case, both of the flows q_{a_k} and $q_{out\,k}$ must fulfil the respective physical constraints

$$\underline{q}_a \leq q_{a_k} \leq \overline{q}_a, \tag{5.18}$$
$$\underline{q}_{out} \leq q_{out\,k} \leq \overline{q}_{out}, \tag{5.19}$$

where $\underline{q}_a$ and $\underline{q}_{out}$ are the minimum flows through sewer mains q_a and q_{out}, respectively. The values of these minimum flows are assumed to be zero in the nominal configuration (no faults). For this case, the MLD form of the element includes the constraints (5.18)–(5.19) and corresponds to a system model that truncates input signal values that differ from their nominal values.
The expressions for the dynamics in MLD form with this assumption are obtained as follows. The given input value $q_a^\star$ must be smaller than the inflow to the element. Otherwise, $q_a^\star$ does not make sense, as the difference $q_a^\star - q_{in_k}$ corresponds to the entry of nonexistent sewage into the element. This idea is mathematically expressed as

$$q_{a1\,k} = \begin{cases} q_a^\star & \text{if} \quad q_a^\star \leq q_{in_k}, \\ q_{in_k} & \text{otherwise.} \end{cases} \tag{5.20}$$

The upper bound of the flow capacity for the sewer main adds another condition for $q_a^\star$, which is given by the upper physical bound as

$$q_{a2_k} = \begin{cases} q_{a1\,k} & \text{if} \quad q_{a1\,k} \leq \overline{q}_a, \\ \overline{q}_a & \text{otherwise.} \end{cases} \tag{5.21}$$

Finally, the maximum volume capacity of the tank also constrains the value of $q_a^\star$ to

$$q_{a3_k} = \begin{cases} v_r & \text{if} \quad q_{a2_k} - q_{out\,k} \geq v_r, \\ q_{a2_k} & \text{otherwise.} \end{cases} \tag{5.22}$$

Moreover, the given value $q_{out}^\star$ is also restricted by its own physical constraint in (5.19). The corresponding expression that restricts this flow due to its physical upper bound is

$$q_{out1\,k} = \begin{cases} q_{out}^\star & \text{if} \quad q_{out}^\star \leq \overline{q}_{out}, \\ \overline{q}_{out} & \text{otherwise,} \end{cases} \tag{5.23}$$

Table 5.1 Expressions for the δ and z variables in the RTIG model that considers manipulated flows in a closed-loop simulation

Logical variable δ	Auxiliary variable z
$[\delta_{1k}=1] \longleftrightarrow [q_a^\star \geq q_{ink}]$	$z_{1k}=\delta_{1k}q_a^\star+(1-\delta_{1k})q_{ink}$
$[\delta_{2k}=1] \longleftrightarrow [z_{1k} \leq \overline{q}_a]$	$z_{2k}=\delta_{2k}z_{1k}+(1-\delta_{2k})\overline{q}_a$
$[\delta_{3k}=1] \longleftrightarrow [q_{out}^\star \leq \overline{q}_{out}]$	$z_{3k}=\delta_{3k}q_{out}^\star+(1-\delta_{3k})\overline{q}_{out}$
$[\delta_{4k}=1] \longleftrightarrow [z_{3k} \geq \beta v_k]$	$z_{4k}=\delta_{4k}z_{3k}+(1-\delta_{4k})\beta v_k$
$[\delta_{5k}=1] \longleftrightarrow [z_{2k}-z_{4k} \geq v_{rk}]$	$z_{5k}=\delta_{5k}v_{rk}+(1-\delta_{5k})z_{2k}$

and so the expression that restricts this flow due to the actual tank volume is written as

$$q_{out2k}=\begin{cases} q_{out1k} & \text{if} \quad q_{out1k} \geq \beta v_k, \\ \beta v_k & \text{otherwise.} \end{cases} \tag{5.24}$$

Again, the flow q_{bk} is defined by the mass conservation at the input gate as $q_{bk}=q_{ink}-q_{a3k}$.

The MLD model for this case is then obtained from the expressions presented in Table 5.1, which define the corresponding δ and z variables. Here, $q_{bk}=q_{ink}-z_{5k}$.

The difference equation associated with the tank dynamics is written as

$$v_{k+1}=v_k+\Delta t\left(z_{5k}-z_{4k}\right). \tag{5.25}$$

Finally, the MLD form (2.8), taking the volume as the system output, is written as

$$v_{k+1}=v_k+\begin{bmatrix}0 & 0 & 0 & -\Delta t & \Delta t\end{bmatrix}\begin{bmatrix} z_{1k} \\ z_{2k} \\ z_{3k} \\ z_{4k} \\ z_{5k} \end{bmatrix}, \tag{5.26a}$$

$$y_k=v_k, \tag{5.26b}$$

and inequality (2.8c), which actually collects together 30 linear inequalities that were automatically generated by HYSDEL.

2. The RTIG element is used to build a global hybrid model of the system for prediction purposes within a closed-loop control scheme. In this case, constraints (5.18) and (5.19) are included in the controller design, so they are not taken into account when the MLD model of the element is obtained. Thus, the expressions for the element dynamics can be obtained as follows. In order to restrict the value of the given input $q_a^\star$ so as to fulfil mass conservation at the input gate, the flow through link q_a is expressed as

Table 5.2 Expressions for the δ and z variables in the RTIG model that considers manipulated flows in closed-loop prediction

Logical variable δ	Auxiliary variable z
$[\delta_{1k} = 1] \longleftrightarrow [q_a^\star \leq q_{in\,k}]$	$z_{1k} = \delta_{1k} q_a^\star + (1 - \delta_{1k}) q_{in\,k}$
$[\delta_{2k} = 1] \longleftrightarrow \left[z_{1k} - z_{3k} \leq \frac{\overline{v} - v_k}{\Delta t}\right]$	$z_{2k} = \delta_{2k} q_{a1\,k} + (1 - \delta_{2k}) \frac{\overline{v} - v_k}{\Delta t}$
$[\delta_{3k} = 1] \longleftrightarrow [q_{out}^\star \leq \beta v_k]$	$z_{3k} = \delta_{3k} q_{out}^\star + (1 - \delta_{3k}) \beta v_k$

$$q_{a1\,k} = \begin{cases} q_a^\star & \text{if} \quad q_a^\star \leq q_{in\,k}, \\ q_{in\,k} & \text{otherwise.} \end{cases} \tag{5.27}$$

However, the volume capacity of the tank also restricts the inflow according to the expression

$$q_{a\,k} = \begin{cases} q_{a1\,k} & \text{if} \quad q_{a1\,k} - q_{out\,k} \leq \frac{\overline{v} - v_k}{\Delta t}, \\ \frac{\overline{v} - v_k}{\Delta t} & \text{otherwise.} \end{cases} \tag{5.28}$$

With regards to the tank outflow, the given input $q_{out}^\star$ is restricted according to the outflow relating to the current volume. That is,

$$q_{out\,k} = \begin{cases} q_{out}^\star & \text{if} \quad q_{out}^\star \leq \beta v_k, \\ \beta v_k & \text{otherwise.} \end{cases} \tag{5.29}$$

The expressions for δ and z that are used to obtain the corresponding MLD model are collected in Table 5.2.

The MLD form (2.8) for this element in this case, taking the tank volume as the system output, is written as follows:

$$v_{k+1} = v_k + \begin{bmatrix} 0 & -\Delta t & \Delta t \end{bmatrix} \begin{bmatrix} z_{1k} \\ z_{2k} \\ z_{3k} \end{bmatrix}, \tag{5.30a}$$

$$y_k = v_k, \tag{5.30b}$$

and inequality (2.8c), which actually collects together 18 linear inequalities that were automatically generated by HYSDEL.

5.1.3 Redirection Gates (RG)

These elements are used to redirect the sewage flow to a certain point in the network. Assuming that q_{a_k} is manipulated, outflow q_{b_k} in Figure 5.3 must fulfil the mass conservation law at this point. Generally, q_{a_k} is assumed to be limited, in which case q_{b_k} is unlimited. The reason for this assumption is

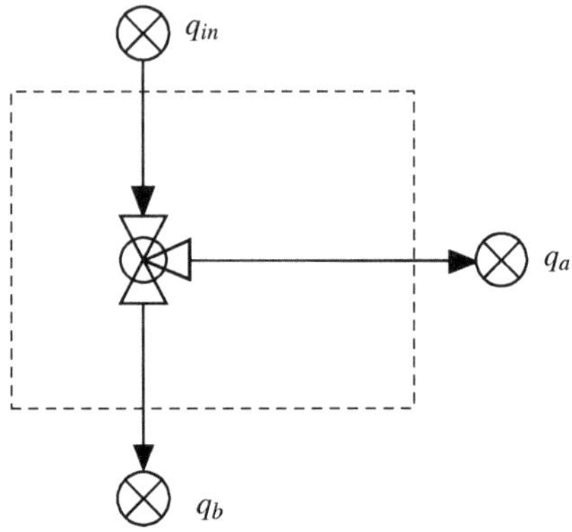

Figure 5.3 Scheme of a redirection gate element

that, if both outflows are limited, the situation where q_{in_k} would be larger than the sum of these limits could occur, causing the whole optimisation problem associated with MPC design to have no feasible solution. The expressions that describe the element dynamics are as follows:

$$q_{a_k} = \begin{cases} q_{in_k} & \text{if} \quad q_{in_k} \leq \overline{q}_a, \\ \overline{q}_a & \text{otherwise}, \end{cases} \tag{5.31a}$$

$$q_{b_k} = q_{in_k} - q_{a_k}. \tag{5.31b}$$

If the flow through sewer main q_a is imposed (*e.g.*, computed using a given control law), its new expression is

$$q_{a_k} = \begin{cases} \overline{q}_a & \text{if} \quad q_a^\star > \overline{q}_a, \\ q_a^\star & \text{otherwise}, \end{cases} \tag{5.32}$$

where $q_a^\star$ corresponds to the imposed/computed value for the flow q_{a_k}, while q_{b_k} follows (5.31b).

The MLD model of this element depends on whether the sewer main q_a is manipulated. In any case, this is an static element; *i.e.*, a state variable is not defined in the MLD model. So, the element only adds more δ and z variables and a set of linear inequalities to the global MLD model of the sewer network.

q_a as an Unmanipulated Flow

This model is basically used for open-loop simulation. The hybrid dynamics are defined by the maximum flow through sewer main q_a, which causes the definition of the auxiliary logic variable

$$[\delta_k = 1] \longleftrightarrow [q_{in\,k} \geq \overline{q}_a], \tag{5.33}$$

and the redefinition of flow $q_{a\,k}$ as

$$\begin{aligned} z_k &= q_{a\,k}, \\ &= \delta_k \overline{q}_{a\,k} + (1-\delta_k)\, q_{in\,k}. \end{aligned} \tag{5.34}$$

Thus, the flow through the sewer main q_b is defined by mass conservation as

$$q_{b\,k} = q_{in\,k} - z_k. \tag{5.35}$$

In this case, six linear inequalities are defined, as in (2.8c). They are expressed as follows:

$$\begin{bmatrix} -\overline{q}_a \\ \overline{q}_{in} - \overline{q}_a \\ \overline{q}_{in} - \overline{q}_a \\ \overline{q}_a \\ -\overline{q}_a \\ -\overline{q}_{in} + \overline{q}_a \end{bmatrix} \delta_k + \begin{bmatrix} 0 \\ 0 \\ -1 \\ 1 \\ -1 \\ 1 \end{bmatrix} z_k \leq \begin{bmatrix} 1 \\ -1 \\ -1 \\ 1 \\ 0 \\ 0 \end{bmatrix} q_{in\,k} + \begin{bmatrix} \overline{q}_a \\ \overline{q}_{in} \\ \overline{q}_{in} - \overline{q}_a \\ \overline{q}_a \\ -\overline{q}_a \\ \overline{q}_a \end{bmatrix}. \tag{5.36}$$

Note that it is stated that the sewer main q_a is the natural path followed by the inflow.

q_a as a Manipulated Flow

In this case, a closed-loop control scheme is considered, so the element could be used for either a simulation or a prediction model. In the first case, the hybrid dynamics are defined while taking into account the system constraint $\underline{q}_a \leq q_a \leq \overline{q}_a$. This constraint is assumed to be added to the global model when the control law is designed. Thus, a couple of δ and z variables are defined in order to check that the given value of q_a fulfils the mentioned constraint. The definitions of these δ variables that depend on the hybrid condition are explained as follows. For the condition of sewage sufficiency, the logical variable δ_1 is determined as

$$[\delta_{1\,k} = 1] \longleftrightarrow [q_{a\,k} \geq q_{in\,k}], \tag{5.37}$$

and for the condition related to the upper limit of sewer main q_a, the logical variable δ_2 is determined as

$$[\delta_{2\,k} = 1] \longleftrightarrow [q_{a\,k} \leq \overline{q}_a]. \tag{5.38}$$

The corresponding auxiliary continuous variables are then

$$z_{1k} = \delta_{1k} q_{ak} + (1-\delta_{1k}) q_{ink}, \tag{5.39}$$

$$z_{2k} = \delta_{2k} z_{1k} + (1-\delta_{2k}) \overline{q}_{ak}. \tag{5.40}$$

Again, the flow through sewer q_b is immediately defined by mass conservation as $q_{bk} = q_{ink} - z_{2k}$. Here, 12 linear inequalities are defined according to inequality (2.8c), with

$$\begin{aligned}
E_1 &= \begin{bmatrix} 0\,0\,0\,0 & -1 & 1 & -1 & 1 & 1 & -1 & 0 & 0 \\ 0\,0\,0\,0 & 0 & 0 & 0 & 0 & -1 & 1 & -1 & 1 \end{bmatrix}^T, \\
E_{21} &= \begin{bmatrix} -\overline{q}_a & \overline{q}_{in} & 0 & 0 & \overline{q}_a & -\overline{q}_{in} \\ 0 & 0 & -\overline{q}_a & 0 & 0 & 0 \end{bmatrix}, \\
E_{22} &= \begin{bmatrix} -\overline{q}_{in} & \overline{q}_a & 0 & 0 & 0 & 0 \\ 0 & 0 & (\overline{q}_{in} - \overline{q}_a) & \overline{q}_a & -\overline{q}_a & -(\overline{q}_{in} - \overline{q}_a) \end{bmatrix}, \\
E_2 &= \begin{bmatrix} E_{21} \; E_{22} \end{bmatrix}^T, \\
E_3 &= \begin{bmatrix} 0\,0\,0\,0 & -1 & 1 & -1 & 1 & 1 & -1 & 0 & 0 \\ 0\,0\,0\,0 & 0 & 0 & 0 & 0 & -1 & 1 & -1 & 1 \end{bmatrix}^T, \\
E_5 &= \begin{bmatrix} 0 \; \overline{q}_{in} \; -\overline{q}_a \; \overline{q}_a \; \overline{q}_a \; \overline{q}_{in} \; 0 \; 0 \; (\overline{q}_{in} - \overline{q}_a) \; \overline{q}_a \; -\overline{q}_a \; \overline{q}_a \end{bmatrix}^T.
\end{aligned} \tag{5.41}$$

On the other hand, when a prediction model is considered, q_a theoretically fulfils the design constraint $\underline{q}_a \leq q_a \leq \overline{q}_a$, but the values of $\underline{q}_a$ and $\overline{q}_a$ are information for the control algorithm, so they could be different from the current physical bounds (this could occur due to the effect of a fault, for example). Hence, the definitions

$$[\delta_k = 1] \longleftrightarrow [q_{ak} \geq q_{ink}] \tag{5.42}$$

and

$$z_k = \delta_{1k} q_{ak} + (1-\delta_{1k}) q_{ink} \tag{5.43}$$

are provided in order to fulfil mass conservation. This element therefore adds the following six inequalities to the global model according to (2.8c):

$$\begin{bmatrix} -\overline{q}_a \\ \overline{q}_{in} \\ \overline{q}_a \\ \overline{q}_{in} \\ -\overline{q}_{in} \\ -\overline{q}_a \end{bmatrix} \delta_k + \begin{bmatrix} 0 \\ 0 \\ -1 \\ 1 \\ -1 \\ 1 \end{bmatrix} z_k \leq \begin{bmatrix} 1 & -1 \\ -1 & 1 \\ 0 & -1 \\ 0 & 1 \\ -1 & 0 \\ 1 & 0 \end{bmatrix} \begin{bmatrix} q_{ink} \\ q_{ak} \end{bmatrix} + \begin{bmatrix} 0 \\ \overline{q}_{in} \\ \overline{q}_a \\ \overline{q}_{in} \\ 0 \\ 0 \end{bmatrix}. \tag{5.44}$$

In conclusion, which of these two models is chosen is directly related to the use of the RG element within the global network model and the use of the model itself.

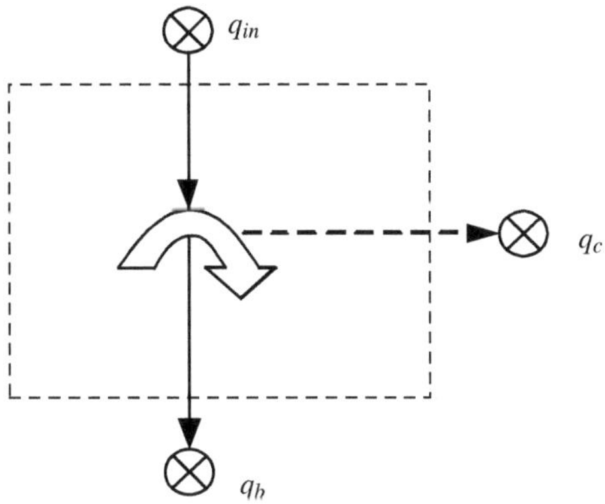

Figure 5.4 Scheme of a sewage pipe

5.1.4 *Sewage Pipes (SP)*

The flow links between tanks have a limited flow capacity. As the flow from virtual tanks cannot be controlled, when this flow limit is exceeded, the resulting overflow may be redirected to a tank to which the original link was not connected. When RGs are used to redirect the virtual tank outflow, the *unmanipulated* link associated with the RG could surpass its maximum sewage flow capacity. Hence, the sewage overflow is sent to a tank located downstream, or it flows into the receiving environment as pollution. This behaviour can be represented by the following equations (see Figure 5.4):

$$q_{bk} = \begin{cases} \overline{q}_b & \text{if} \quad q_{in} > \overline{q}_b, \\ q_{ink} & \text{otherwise}, \end{cases} \tag{5.45}$$

$$q_{ck} = \begin{cases} q_{ink} - \overline{q}_b & \text{if} \quad q_{in} > \overline{q}_b, \\ 0 & \text{otherwise}, \end{cases} \tag{5.46}$$

where $\overline{q}_b$ is the maximum flow through the sewer main q_b, flow q_{ink} is the element inflow, and q_{ck} corresponds to the outflow.

Just one logical variable is needed for the MLD model of this element. It is defined from the hybrid overflow condition as

$$[\delta_k = 1] \longleftrightarrow [q_{in} \geq \overline{q}_b], \tag{5.47}$$

and the auxiliary continuous variables that define the flows q_{bk} and q_{ck} are, respectively,

$$\begin{aligned} z_{1k} &= q_{bk}, \\ &= \delta_k \overline{q}_b + (1 - \delta_k)\, q_{ink}, \end{aligned} \tag{5.48a}$$

and

$$\begin{aligned} z_{2k} &= q_{ck}, \\ &= \delta_k \left(q_{in_k} - \overline{q}_b \right). \end{aligned} \tag{5.48b}$$

As in the RG case, note that this is a static element, so it only adds more δ and z variables to a global MLD model of the sewer network, as well as more constraints in (2.8c), as follows:

$$\begin{bmatrix} -\overline{q}_{in}+\overline{q}_b \\ \overline{q}_b \\ \overline{q}_b \\ \overline{q}_{in}-\overline{q}_b \\ -\overline{q}_{in}+\overline{q}_b \\ -\overline{q}_b \\ \overline{q}_{in}-\overline{q}_b \\ \overline{q}_b \\ -\overline{q}_b \\ -\overline{q}_{in}+\overline{q}_b \end{bmatrix} \delta_k + \begin{bmatrix} 0 & 0 \\ 0 & 0 \\ -1 & 0 \\ 1 & 0 \\ -1 & 0 \\ 1 & 0 \\ 0 & -1 \\ 0 & 1 \\ 0 & -1 \\ 0 & 1 \end{bmatrix} z_k \leq \begin{bmatrix} -1 \\ 1 \\ 0 \\ 0 \\ -1 \\ 1 \\ -1 \\ 1 \\ 0 \\ 0 \end{bmatrix} q_{in_k} + \begin{bmatrix} \overline{q}_b \\ 0 \\ 0 \\ \overline{q}_{in} \\ 0 \\ 0 \\ \overline{q}_{in} \\ 0 \\ 0 \\ 0 \end{bmatrix}, \tag{5.49}$$

where $\overline{q}_{in}$ corresponds to the maximum inflow to the element.

Overflows in VTs and SPs do not necessarily flow towards the nearest element downstream. In general, SP overflows can flow to either VTs or a receiving environment (as pollution).

5.1.5 The Entire MLD Catchment Model

It is possible to obtain the global MLD model by connecting all of the elements in order to build the equivalent model scheme for the BTC, as shown in Figure 5.5. However, a software tool that allows the corresponding inputs and outputs of each element to be interpreted in its MLD form, and which translates them into a global MLD is now pending implementation. The final tool would have the advantages of an existing tool for sewer networks, CORAL (Figueras *et al.*, 2002), but it would include the hybrid model framework in its kernel.

The manipulated variables of the system, denoted q_u, are the manipulated variables of each component, as described previously. The whole sewer network expressed in MLD form can be written as

$$v_{k+1} = Av_k + B_1 q_{u_k} + B_2 \delta_k + B_3 z_k + B_4 d_k, \tag{5.50}$$

$$y_k = Cv_k + D_1 q_{u_k} + D_2 \delta_k + D_3 z_k + D_4 d_k, \tag{5.51}$$

$$E_2 \delta_k + E_3 z_k \leq E_1 q_{u_k} + E_4 v_k + E_5 + E_6 d_k, \tag{5.52}$$

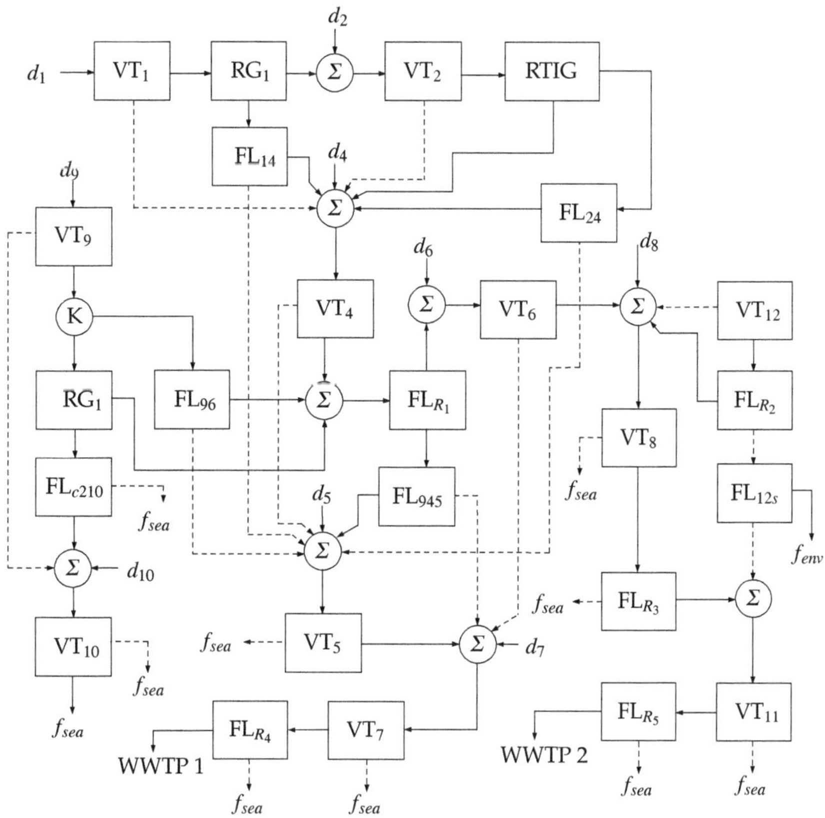

Figure 5.5 BTC scheme using hybrid network elements

where $v \in \mathbb{V} \subseteq \mathbb{R}_+^{n_c}$ corresponds to the vector of tank volumes (states), $q_u \in \mathbb{U} \subseteq \mathbb{R}_+^{mi_c}$ is the vector of manipulated sewage flows (inputs), $d \in \mathbb{R}_+^{md}$ is the vector of rain measurements (disturbance), the logic vector $\delta \in \{0,1\}^{r_\ell}$ collects together the Boolean overflow conditions, and the vector $z \in \mathbb{R}_+^{r_c}$ is associated with variables that may or may not appear depending on the system states and inputs. Variables δ and z are auxiliary variables associated with the MLD form. Inequality (5.52) collects the set of system constraints as well as translations from logic propositions. Notice that (5.50) is a more general MLD than the one presented in Bemporad and Morari (1999), due to the addition of the measured disturbances.

Likewise, measured disturbances may be considered inputs or system states. Depending on the case, q_u and d could be collected in a single vector of system inputs that would depend on the selected control algorithm. Hence, the MLD form could be rewritten as

$$v_{k+1} = Av_k + \tilde{B}\tilde{q}_{uk} + \begin{bmatrix} B_2 & B_3 \end{bmatrix} \begin{bmatrix} \delta_k \\ z_k \end{bmatrix}, \tag{5.53a}$$

$$y_k = Cv_k + \tilde{D}\tilde{q}_{uk} + \begin{bmatrix} D_2 & D_3 \end{bmatrix} \begin{bmatrix} \delta_k \\ z_k \end{bmatrix}, \tag{5.53b}$$

$$\begin{bmatrix} E_2 & E_3 \end{bmatrix} \begin{bmatrix} \delta_k \\ z_k \end{bmatrix} \leq \tilde{E}\tilde{q}_{uk} + \begin{bmatrix} E_4 & E_5 \end{bmatrix} \begin{bmatrix} v_k \\ 1 \end{bmatrix}, \tag{5.53c}$$

where the vector $\tilde{q}_{uk}$ collects the control inputs and measured disturbances. Moreover,

$$\tilde{B} = [B_1 \quad B_4], \quad \tilde{D} = [D_1 \quad D_4], \quad \text{and} \quad \tilde{E} = [E_1 \quad E_6].$$

On the other hand, assuming that the rain prediction over H_p obeys pre-established linear dynamics that can be described via

$$d_{k+1} = A_d d_k, \tag{5.54}$$

the MDL form in (5.50) can be rewritten as

$$\begin{bmatrix} v_{k+1} \\ d_{k+1} \end{bmatrix} = \begin{bmatrix} A & B_4 \\ 0 & A_d \end{bmatrix} \begin{bmatrix} v_k \\ d_k \end{bmatrix} + \begin{bmatrix} B_1 & B_2 & B_3 \\ 0 & 0 & 0 \end{bmatrix} \begin{bmatrix} q_{uk} \\ \delta_k \\ z_k \end{bmatrix}, \tag{5.55a}$$

$$y_k = \begin{bmatrix} C & 0 \end{bmatrix} \begin{bmatrix} v_k \\ d_k \end{bmatrix} + \begin{bmatrix} D_1 & D_2 & D_3 \end{bmatrix} \begin{bmatrix} q_{uk} \\ \delta_k \\ z_k \end{bmatrix}, \tag{5.55b}$$

$$\begin{bmatrix} E_2 & E_3 \end{bmatrix} \begin{bmatrix} \delta_k \\ z_k \end{bmatrix} \leq \begin{bmatrix} E_4 & E_6 \end{bmatrix} \begin{bmatrix} v_k \\ d_k \end{bmatrix} + \begin{bmatrix} E_1 & E_5 \end{bmatrix} \begin{bmatrix} q_{uk} \\ 1 \end{bmatrix}. \tag{5.55c}$$

In general, this assumption about the disturbance dynamics is an open research topic. Different types of rain prediction can be considered, since this procedure can be developed in either a theoretical way, such as by statistical approaches, AR models, *etc.* (Smith and Austin, 2000; Zaw and Naing, 2008; Zhu *et al.*, 2008; Karamouz *et al.*, 2009), or in a practical way, using for example radar, meteorological satellites, *etc.* (Yuan *et al.*, 1999; Thielen *et al.*, 2000; Katayama *et al.*, 2006; Mandapaka *et al.*, 2009). According to Cembrano and Quevedo (1999), different assumptions can be made for rain prediction when a receding horizon control is used in the RTC of sewer networks. Results show that the assumption of constant rain over a short prediction horizon gives results that are comparable with those obtained under the assumption of known rain over the considered horizon, confirming similar results reported in Gelormino and Ricker (1994) and Ocampo-Martinez *et al.* (2005). Based on this, matrix A_d in (5.54) can be set as an identity matrix of suitable dimensions.

As will be discussed in Chapter 9, this topic is an open research field, and detailed study in this area should yield results that will enable the realisation of more accurate system prediction models, thus resulting in better MPC controller designs.

5.2 Predictive Control Strategy

This section discusses the design of MPC controllers for sewer networks based on hybrid system models. The different aspects discussed here are applied to the particular case study of this book, but they are also easily extrapolated to other sewage system topologies. The concepts and definitions of Section 2.2 are applied in a straightforward manner in this section, but the particular notation used for sewer networks is employed. Hence, the sequences related to manipulated flows and tank volumes are defined as

$$\mathbf{v}_k = \{v_{1|k}, v_{2|k}, \ldots, v_{H_p|k}\}, \tag{5.56a}$$

$$\mathbf{q_u}_k = \{q_{u0|k}, q_{u1|k}, \ldots, q_{uH_p-1|k}\}, \tag{5.56b}$$

and the expression for an admissible input sequence with respect to the initial volume $v_{0|k} \triangleq v_k \in \mathbb{V}$ is now written as

$$\mathscr{Q}_{\mathscr{U}}(v_k) \triangleq \{\mathbf{q_u}_k \in \mathbb{U}^{H_p} | \mathbf{v}_k \in \mathbb{V}^{H_p}\}. \tag{5.57}$$

5.2.1 *Control Objectives*

In the design of an MPC controller based on a hybrid system model, the control objectives are the same as those in the controller design in Chapter 4 (Section 4.2.1). According to the hybrid model for sewer networks proposed in Section 5.1, all overflows and flows to WWTPs are defined by auxiliary variables z. However, they can also be defined as system outputs: either individual (over)flows or sums of (over)flows, depending on the particular case.

5.2.2 *Cost Function*

Each control objective defines or can define one term in the cost function. Hence, the expression of that function depends on its constitutive variables (auxiliary or output type). In particular, the structure of the cost function (2.14a) takes the form

$$J(\mathbf{q_u}_k, \Delta_k, \mathbf{z}_k, v_k) \triangleq \sum_{i=0}^{H_p-1} \left\| Q_z(z_{k+i|k} - z_r) \right\|_p + \sum_{i=0}^{H_p-1} \left\| Q_y(y_{k+i|k} - y_r) \right\|_p, \quad (5.58)$$

where Q_z and Q_y correspond to weight matrices of suitable dimensions that fulfil the conditions in (2.15), and z_r and y_r are reference trajectories that relate to auxiliary and output variables, respectively. For the first and second objectives, the references are no flow. For the third objective, the references are the maximum inflow capacities of the associated WWTPs (flows directly related to the sewage volume that can be treated). Priorities are set by selecting matrices Q_z and Q_y. The norm p can be selected as $p = 1, 2$ or $p = \infty$. Note that, since all of the performance variables are positive, the case where $p = 1$ is actually a simple sum of the performance variables.

According to the definition of the control objectives given in Section 5.1 and the complementary fifth objective, the cost function adopts the form

$$J(\mathbf{q_u}_k, \Delta_k, \mathbf{z}_k, v_k) \triangleq \|Q_{v_f} v_{rt}(H_p|k)\|_p + \sum_{i=0}^{H_p-1} \left\| Q_z\left(z_{k+i|k} - z_r\right) \right\|_p, \quad (5.59)$$

where Q_{v_f} is the corresponding weight matrix of suitable dimensions.

5.2.3 *Problem Constraints*

The modelling approach is based on mass conservation. Therefore, physical restrictions must be included as constraints in the optimisation problem. The sum of the inflows into a node that connects links must equal the outflow. The control variables are limited to the range given in (3.8). The constraints associated with the MIPC problem are in general the constraints associated with hybrid behaviour as well as the system's physical constraints for manipulated flows and real tanks, while the initial condition corresponds to the tank volume measurements at time instant $k \in \mathbb{Z}_+$.

All of the constraints can be expressed in the form given by (5.52). The physical constraints are considered *hard constraints* in the control problem. On the other hand, the overflows from the sewer mains and virtual tanks are considered *soft constraints*, so a constraint manager can be designed and implemented to solve the control problem with constraint prioritisation (Kerrigan *et al.*, 2000; Camacho and Bordons, 2004).

5.2.4 *MIPC Problem*

According to the aspects described previously, the predictive control problem for a sewer network with a hybrid model is defined as the OOP

$$\min_{\mathbf{q_u}_k \in \mathcal{Q}_\mathcal{U}(v_k), \Delta_k, \mathbf{z}_k} \quad J(\mathbf{q_u}_k, \Delta_k, \mathbf{z}_k, v_k), \tag{5.60a}$$

subject to

$$v_{k+i+1|k} = A\, v_{k+i|k} + B_1\, q_{u_{k+i|k}} + B_2\, \delta_{k+i|k} + B_3\, z_{k+i|k} + B_4\, d_{k+i|k}, \tag{5.60b}$$

$$y_{k+i|k} = C\, v_{k+i|k} + D_1\, q_{u_{k+i|k}} + D_2\, \delta_{k+i|k} + D_3\, z_{k+i|k} + D_4\, d_{k+i|k}, \tag{5.60c}$$

$$E_2\, \delta_{k+i|k} + E_3\, z_{k+i|k} \leq E_1\, q_{u_{k+i|k}} + E_4\, v_{k+i|k} + E_5 + E_6\, d_{k+i|k}, \tag{5.60d}$$

$$d_{k+i+1|k} = A_d\, d_{k+i|k}, \tag{5.60e}$$

for $i = 0, \ldots, H_p - 1$. Assuming that the problem is feasible for $v \in \mathbb{V}$, in other words $\mathcal{Q}_\mathcal{U}(v) \neq \emptyset$, the receding horizon philosophy is then used, with

$$q_{u_{\text{MPC}}}(v_k) \triangleq q^*_{u_{0|k}} \tag{5.61}$$

considered to be the MPC law, and the entire optimisation process is repeated at time $k+1$.

5.3 Simulation and Results

5.3.1 *Preliminaries*

The purpose of this section is to show the performance of HMPC for realistic rain episodes. The assumptions made for this implementation will be presented and their validity discussed before the results are given.

The transformation of the hybrid system equations into the MLD form requires the application of the set of rules in (5.1) and (5.2). The higher level language and associated compiler HYSDEL was used to avoid the tedious procedure of deriving the MLD form by hand. Given the MLD model, the controllers were designed and the scenarios simulated using the Hybrid Toolbox for MATLAB® (see Bemporad, 2006). ILOG CPLEX 9.1 was used to solve MIP problems.

The system considered is shown in Figure 5.6. The dashed lines represent the overflows from tanks and sewer mains. These lines therefore represent the hybrid behaviour of the network to some degree. The hybrid model of the BTC has twelve state variables, corresponding to the volumes in the tanks (eleven virtual and one real), four control signals relating to the manipulated flows in gates (three redirection gates and one retention gate), and eleven perturbation signals relating to the rain inflows of each virtual tank. The nominal operating ranges of the control signals, descriptions of the variables shown in Figure 5.6, and all of the parameters needed are provided in

Table 5.3 Relations between z variables and control objectives

Objective	z-Vector	z-Variable	Description
1	z_{str_v}	z_2	Overflow in T_1
		z_6	Overflow in T_2
		z_{10}	Overflow in T_4
		z_{12}	Overflow in T_9
		z_{20}	Overflow in T_5
		z_{22}	Overflow in T_6
		z_{24}	Overflow in T_7
		z_{26}	Overflow in T_{12}
		z_{32}	Overflow in T_8
		z_{36}	Overflow in T_{10}
		z_{40}	Overflow in T_{11}
2	z_{str_q}	z_4	Overflow in q_{14}
		z_8	Overflow in q_{24}
		z_{14}	Overflow in q_{96}
		z_{18}	Overflow in q_{945}
		z_{30}	Overflow in q_{12s}
		z_{34}	Overflow in q_{c210}
3	z_{sea}	z_{29}	Flow to environment (q_{12s})
		z_{35}	Flow to sea (q_{10M})
		z_{38}	Flow to sea (q_{8M})
		z_{42}	Flow to sea (q_{11M})
		z_{44}	Flow to sea (q_{7M})
4	—	z_{43}	Flow to Llobregat WWTP
	—	z_{41}	Flow to Besòs WWTP

Tables 3.1 and 3.2. Table 5.3 collects the auxiliary variables z defined for the BTC and relates them to the control objectives discussed in Section 4.2.1.

In order to define the outputs of the global MLD model of the network, the following variables are defined:

$$y_1 = \sum_i z_{\mathrm{str}_v}, \quad y_2 = \sum_i z_{\mathrm{str}_q}, \quad y_3 = z_{41}, \quad \text{and} \quad y_4 = z_{43}.$$

5.3.2 MLD Model Descriptions and Controller Set-up

Two different MLD models are needed to simulate the scenarios: one for the HMPC controller, MLD_C, and one to simulate the plant, MLD_P. Note that physical constraints are included in the model MLD_C, and the solution to the optimisation problem respects these constraints in the nominal case when there is no mismatch between the model and the plant. Otherwise, the solution to the optimisation problem may not respect the physical

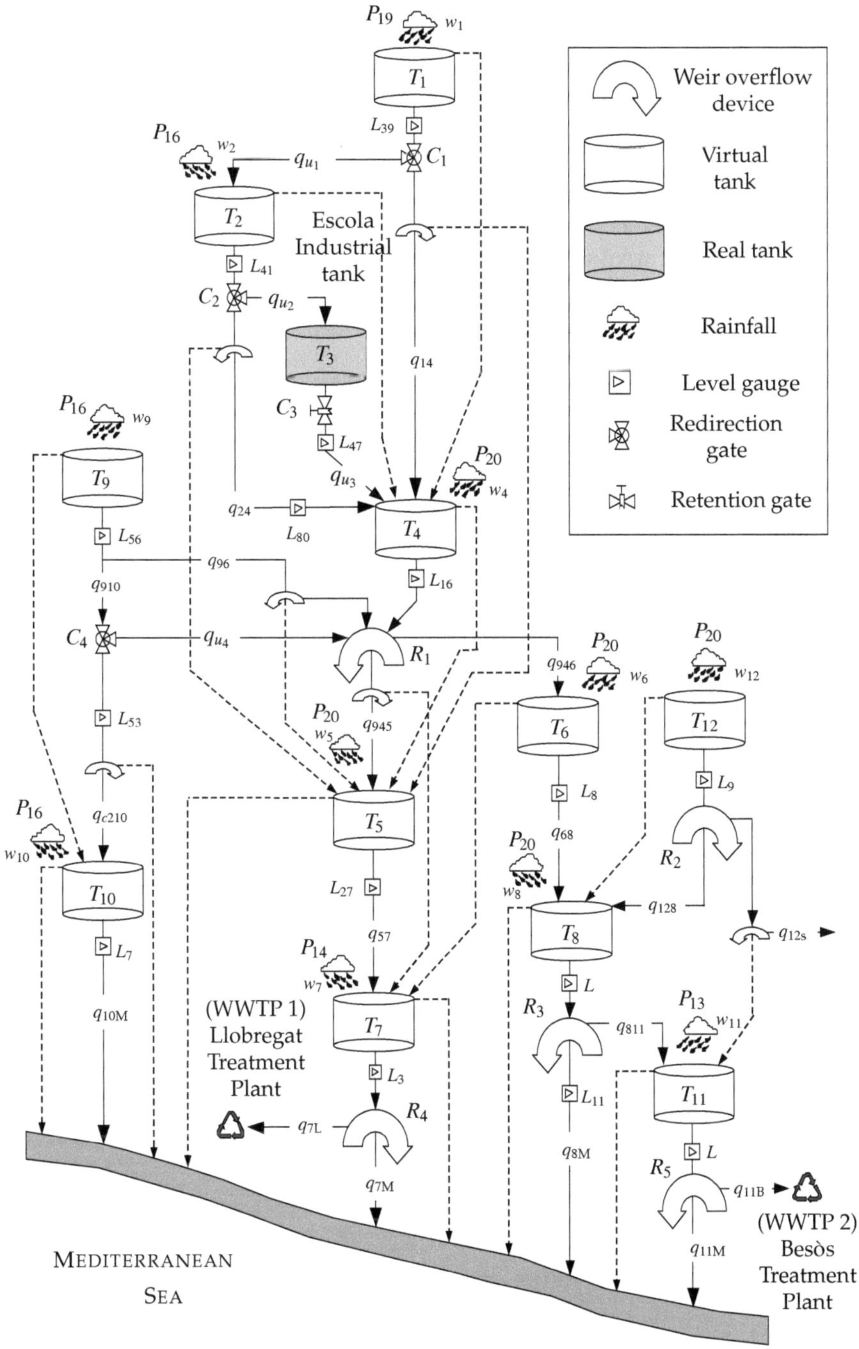

Figure 5.6 BTC diagram for HMPC design

restrictions of the network. The model MLD_P is therefore augmented such that the control signals from the controller are adjusted to respect the physical restrictions of the whole network. For this reason, MLD_P contains more auxiliary variables.

An alternative is to use the set of linear and nonlinear equations related to each network compositional element as the plant model. This implies that MLD_P is not needed. For the simulations and results presented next, this second alternative is used. The implemented model MLD_C has 22 Boolean variables and 44 auxiliary variables, which implies that for each time instant $k \in Z_+$, and considering the prediction horizon $H_p = 6$, $2^{22\times 6} = 5.4 \times 10^{39}$ LP problems (for $p = 1, \infty$) or QP problems (for $p = 2$) need to be solved in the worst case.

In this case of control design where a hybrid model of the system is taken into account, tuning techniques such as weight-based approaches are usually implemented. It has been suggested that other methods, such as the lexicographic approach presented in Chapter 4 for linear MPC, should not be applied to the predictive control of hybrid systems due to the complexity of the optimisation problem for a large-scale networked system. Note that, if the lexicographic approach is implemented, the number of discrete optimisation problems to be solved at each time instant k of the scenario would be equal to the number of control objectives. This would probably result in a high computational burden and complex solving algorithms. Chapter 6 deals with these problems and proposes some possible solutions.

Hence, for this case, where tuning is performed using the weight-based approach, the weight matrices in the cost function (5.58) are given by

$$Q_y = \begin{cases} \text{diag}(w_{\text{str}_v} I_n \quad w_{\text{str}_q} I_n \quad w_{\text{WWTP}} I_n) & \text{if} \quad \text{only } y \text{ are used in } J, \\ w_{\text{WWTP}} I_n & \text{if} \quad z \text{ and } y \text{ are used in } J, \end{cases} \tag{5.62a}$$

and

$$Q_z = \begin{cases} \text{diag}(w_{\text{str}_v} I_n \quad w_{\text{str}_q} I_n) & \text{if} \quad z \text{ and } y \text{ are used in } J, \\ 0 & \text{otherwise}, \end{cases} \tag{5.62b}$$

where the descriptions of the weight parameters w_i are collected in Table 5.4 and I_n corresponds to a identity matrix of suitable dimensions. Moreover, the vector of references is $z_r \triangleq \mathbb{O}_n$, and y_r is set as follows:

$$y_r = \begin{cases} [\mathbb{O}_n \quad \mathbb{O}_n \quad \mathbb{O}_n \quad \mathbb{1}_n \overline{q}_{7\text{L}} \quad \mathbb{1}_n \overline{q}_{11\text{B}}]^T & \text{if} \quad \text{only } y \text{ are used in } J, \\ [\mathbb{O}_n \quad \mathbb{1}_n \overline{q}_{7\text{L}} \quad \mathbb{1}_n \overline{q}_{11\text{B}}]^T & \text{if} \quad z \text{ and } y \text{ are used in } J, \end{cases} \tag{5.63}$$

where $\mathbb{O}_n$ is a vector of zeros and $\mathbb{1}_n$ is a vector of ones, both of which have suitable dimensions for each set of variables related to each control objective.

Table 5.4 Parameters description for HMPC controller tuning

Parameter	Description
w_{str_v}	Weight for tank overflow to the street
w_{str_q}	Weight for link overflow to the street
w_{WWTP}	Weight for overflow to treatment plants

Table 5.5 Closed-loop performance obtained when the rain episode for September 14, 1999 was used

Norm	Variables in J	Tuning w_{str_v}	Flooding (hm^3)	Pollution (hm^3)	TS (hm^3)
∞	Only y	10	84.1	225.3	279.4
∞	Only y	0.1	84.2	225.3	279.4
∞	y and z	100	100.9	225.9	278.7
∞	y and z	0.1	103.2	225.6	279.1
2	y and z	1	94.3	228.3	276.1
2	Only y	0.01	92.8	223.5	280.8

As mentioned before, H_p is set to 6, which is equivalent to 30 min (with the sampling time $\Delta t = 300\,\mathrm{s}$). The optimal solutions are computed for a bounded time interval $k \in [0, 100]$, which implies around 8 h of computational time. These computational times refer to a MATLAB® implementation running on a Intel® Pentium® M 1.73 GHz machine.

5.3.3 *Performance Improvement*

The performance of the control scheme is compared with the simulation of the sewer network without control when the sewer mains associated with the manipulated flows are used as passive elements; *i.e.*, the flows $q_{u_1 k}$, $q_{u_2 k}$ and $q_{u_4 k}$ only depend on the inflow to the corresponding gate and these gates are not manipulated (see Section 5.1), while $q_{u_3 k}$ corresponds to the natural outflow[2] from the real tank, as given by (3.3). Control tuning is performed while taking into account the prioritisation of the control objectives.

In a preliminary study, different norms, cost function structures and cost function weights w_i were tested. In order to assign a hierarchical priority to the control objectives, the w_i values differed by an order of magnitude between objectives. Table 5.5 summarises the results obtained for a heavy rain episode that occurred on September 14, 1999 (see Figure 3.7 (a)).

[2] Flow due to gravity.

Table 5.6 Closed-loop performance for ten representative rain episodes (all performance values in hm^3)

Rain episodes	Open loop Flooding	Open loop Pollution	Open loop TS	Closed loop Flooding	Closed loop Pollution	Closed loop TS
99-09-14	108	225.8	278.4	92.9 (14%)	223.5	280.7
02-10-09	116.1	409.8	533.8	97.1 (16%)	398.7	544.9
99-09-03	1	42.3	234.3	0 (100%)	44.3	232.3
02-07-31	160.3	378	324.4	139.7 (13%)	374.6	327.8
99-10-17	0	65.1	288.4	0	58.1 (11%)	295.3
00-09-28	1	104.5	285.3	1	98 (6%)	291.9
98-09-25	0	4.8	399.3	0	4.8	398.8
01-09-22	0	25.5	192.3	0	25	192.4
02-08-01	0	1.2	285.8	0	1.2	285.8
01-04-20	0	35.4	239.5	0	32.3 (9%)	242.5

The use of the ∞-norm for the first and second objectives implies the minimisation of the largest flow onto the street caused by one of the eleven virtual tanks and/or the six sewer mains considered. However, when two or more virtual tanks or sewer mains produce overflows, only the worst case is minimised.

Considering that the system's open-loop performance for this rain episode has a flooding volume about 108,000, a pollution volume of 225,900, and a volume of treated sewage of 278,300 (all in m^3), the improvement in performance ranges from 4.5% to 22.1% for the first objective. The other objectives remain at almost the same values in most cases, fulfilling the prioritisation required of control tuning based on weights. However, it was observed that some simulations could not be performed for some combinations of parameters due to either numerical problems or parameter settings.

Table 5.6 summarises the results for ten of the more representative rain episodes in Barcelona between 1998 and 2002. The results were obtained for the cost function

$$J(u_k, x_k) = \sum_{i=1}^{H_p} \left\| y_{k+i|k} \right\|_{Q_y}^2, \tag{5.64}$$

with $w_{\text{str}_v} = 10^{-2}$. The system performance is enhanced when the hybrid-model-based MPC strategy is applied (see the percentages for some values).

5.4 Summary

The possibility of creating a linear model of a sewer network that takes into account the switching dynamics provided by some constitutive elements was discussed in this chapter. A new modelling methodology for sewer networks using MLD forms was proposed and explained in depth. This approach makes possible to take advantage of the capabilities of MPC when designing a control strategy for the entire system.

The control design was explained and the discrete optimisation problem was described and discussed. The main improvements afforded by the technique and possible problems with its implementation were pointed out. The modelling methodology as well as the controller design were implemented in a simulation of the BTC, and the main results obtained were presented and discussed.

The main drawback with this approach, as highlighted in this chapter, is the real-world practicality of implementing the proposed control design. Some of the simulations presented in Tables 5.5 and 5.6 imply high computational burdens, which makes it impossible to implement RTC based on those controller designs on-line. Thus, an MPC controller based on a hybrid model for a large-scale system can only really be considered for off-line simulation purposes. Chapter 6 deals with this issue of computational burden, proposing and discussing some ways of solving it.

Chapter 6
Suboptimal Hybrid Model Predictive Control

6.1 Motivation

The results obtained in the simulation study of the previous chapter highlighted the significantly improved performance that can be achieved when the HMPC strategy is utilised for the management of sewer networks. Furthermore, the hybrid modelling methodology allows the straightforward treatment of switching phenomena such as overflows and flooding. HMPC has been applied successfully to a variety of control problems in diverse disciplines in recent times, using several approaches (see, among many others, Branicky *et al.*, 1998; De Schutter, 1999; Tomlin *et al.*, 2000; Lincoln and Rantzer, 2001; Bemporad *et al.*, 2002a; Beck *et al.*, 2005; Geyer *et al.*, 2005; Fiacchini *et al.*, 2007; Negenborn *et al.*, 2007; Julius *et al.*, 2007).

However, the underlying optimisation problem of HMPC is combinatorial and $\mathcal{NP}$-hard (Papadimitriou, 1994). The worst-case computation time increases exponentially with the number of combinatorial variables. Figure 6.1 shows how this problem manifests itself in the case study of this book. In Figure 6.1 (a), the rain intensities corresponding to the five rain gauges in the BTC for the critical portion of the rain episode (the second rain peak) on October 17, 1999 are shown. This episode was relatively intense given its return rate of 0.7 years.

In Figure 6.1 (b), the computation time per sample that was required to solve the MIP problem is shown for the same scenario. Recalling that the sampling time of the networked system considered here is 300 s, these plotted results show that the MIP solver is not able to find the optimal solution within this period of time. Furthermore, it is apparent that the computation time varies greatly. The computation time is very small before sample 16.

If optimality is not achieved in a time $t \leq \Delta t$, then at the very least feasibility is required. Often feasibility is sufficient to prove the stability of the MPC scheme (Scokaert *et al.*, 1999). The CPLEX solver used in the current application can be configured to focus on finding a feasible solution before

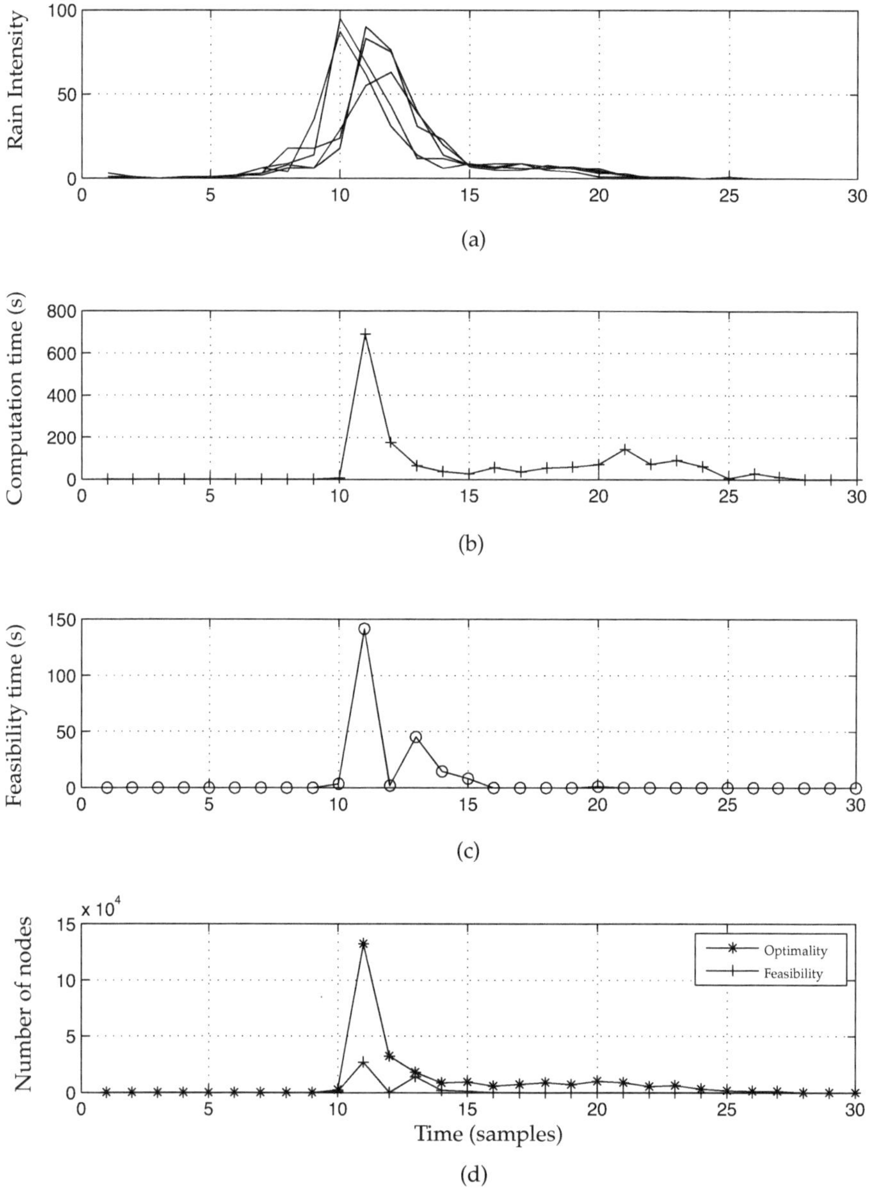

Figure 6.1 Characteristics of the MIP problem for the rain episode that occurred on October 17, 1999: (a) rain intensity, (b) computation time, (c) feasibility time, and (d) number of nodes

an optimal one (ILOG, 2003b). In Figure 6.1 (c), the time needed to find a feasible solution is shown. This was found by iteratively increasing the maximum solution time allowed for the solver until a feasible solution was found. The feature of CPLEX that causes it to search for a feasible solution was activated. Again, it can be seen that the time required to find a feasible solution varies considerably. Furthermore, it should be stated that the feasible solutions found were often of poorer quality than the solutions found by running the system in an open loop.

In the case study considered in this book, the MIQP problem (5.60) is solved for each sample. It takes the following generic form:

$$\min_{\rho} \ \rho^T H \rho + f^T \rho \tag{6.1a}$$

$$\text{subject to } \ A\rho \leq b + Cx_0, \tag{6.1b}$$

where x_0 is the vector that contains the initial conditions of the system and the predicted disturbances (rain). The vector x_0 is the only parameter that changes from sample to sample. The ability of the MIP solver to reduce the computation time from the worst-case computation time depends on its ability to exclude as many nodes as possible from consideration when branching and bounding. This is done by either proving that those nodes are unfeasible or that their solution is suboptimal compared to other solutions. The increase in computational burden is thus linked to the increase in the number of feasible nodes. In Figure 6.1 (d), the number of nodes that the CPLEX solver explored during branching is shown. It is apparent that there was a huge increase in the number of nodes explored between samples. There is thus a dramatic increase in the complexity of the optimisation problem for certain values of x_0.

Physical insight into the process can explain the increase in complexity at time 11. At that time, due to the rain, many of the virtual tanks are very close to overflowing. This in turn means that more trajectories for distinct switching sequences Δ_k are feasible. Similar behaviour was observed in other rain episodes and when other cost functions were used.

As mentioned before, the simulations described in Chapter 5 show the improvement in system performance obtained when an HMPC controller is utilised. However, for some rain episodes, the resulting computation times were quite high compared to the sampling time of the system. This fact shows the extreme randomness of the computation time since the MIP problem greatly depends on the values of the initial conditions for each sample. Table 6.1 presents computation times for particular simulations of the closed-loop system during each rain episode; note that they vary greatly from one simulation to another.

The same problem with computational burden occurs in nonlinear MPC (Findeisen and Allgoer, 2001; Maciejowski, 2002). When the optimisation problem is no longer convex, it must be considered how long the optimisation will take, and whether the quality of the solution will be sufficient to justify the application of the MPC control approach.

Table 6.1 Computation times for simulations of the closed-loop system considering ten representative rain episodes

Rain episodes	Total CPU time (s)	Maximum CPU time per sample (s)
99-09-14	1445.9	1058.3
02-10-09	1328.8	188.4
99-09-03	135.6	60.9
02-07-31	1806.8	391.9
99-10-17	815.7	689.9
00-09-28	254.7	65.6
98-09-25	45.6	40.3
01-09-22	63.3	14.4
02-08-01	47.1	5.6
01-04-20	129.1	18.9

As discussed throughout this book, sewer networks are considered to be large-scale systems. Results in the related research literature indicate that the computation times for such systems are extremely difficult to predict when the associated MIP problem is solved. As the HMPC is based on solving an MIP problem (MILP or MIQP), it is well known that general MIP problems belong to the class of $\mathcal{NP}$-hard problems (Papadimitriou, 1994), and solution algorithms of polynomial complexity do not yet exist (Till *et al.*, 2004).

Note that the large-scale system comprises not only the hybrid model of the sewer network but also the associated MIP problem. Each Boolean variable induces a particular *mode* in the continuous part (Geyer *et al.*, 2003). Therefore, the complexity of this type of large-scale system is related to the number of logical variables in the system model, which yields a large number of possible modes. In an MIP problem, the number of *possible modes* Γ is given by

$$\Gamma = 2^{r_\ell H_p}. \tag{6.2}$$

Therefore, in what follows, a large-scale system is assumed to be one with a large number of Boolean variables and hence a high value of Γ.

Modern MIP solvers such as CPLEX (ILOG, 2003b) include modern branch-and-bound search algorithms that construct a decision tree in steps. The complexity of the tree is given by the total number of decision/logical variables of the system r_ℓ and the time horizon H_p associated with the optimisation problem. At each node, a feasibility test is performed and the cost function value is computed and compared with the lowest upper bound found thus far. If the new cost function value is greater than the lowest upper bound, the corresponding branch is ignored – it is not explored any further. The upper bound is taken from the best integer solution found prior to the actual node (Floudas and Pardalos, 2001). Note that the total number of tree nodes corresponds to the value of Γ. As Γ increases, the computation time required to solve the MIP problem increases.

However, in the worst case, if an MIP solver finds the solution of a large-scale problem by considering all of its possible modes, the computation time would tend to infinity (ILOG, 2003b). Moreover, there are some modes that cannot be reached due to system constraints and the initial conditions of the states. This implies the determination of a subset of Γ that includes the *feasible modes*, defined as a function of the hybrid model equations (MLD form, PWA form, *etc.*), the prediction horizon, and the initial conditions of the system states. Thus, the computation time issue depends in a straightforward manner on the number of feasible modes (see Figure 6.1 (d)).

6.2 General Aspects

6.2.1 Phase Transitions in MIP Problems

MIP solver performance has greatly improved in the last few years (Wojtaszek and Chinneck, 2009). The limit on the size of a problem that is considered to be practically solvable has steadily increased. Part of the reason for this is the steady growth in desktop computing power over the years. However, there has also been tremendous advances in solution algorithms for LPs and QPs, which are a cornerstone of MIP solvers (Bixby *et al.*, 2000; Bixby, 2002; Mancini *et al.*, 2008). Furthermore, modern solvers incorporate many performance-enhancing features that were previously utilised in the literature, such as cutting plane capabilities. The solvers generally apply a barrage of techniques to each particular case. A recent improvement in solving the optimisation problem of HMPC that involves using symbolic techniques to deal with constraint satisfaction problems (CSP) was presented in Bemporad and Giorgetti (2006).

The MIP problem can be considered to be equivalent to the archetypal *$\mathcal{NP}$-complete K-satisfiability* problem (or the *zero-one integer programming* problem, ZIOP) (see Monasson *et al.*, 1999; Blondel and Tsitsiklis, 2000). It was recently shown that K-satisfiability problems exhibit phase transitions in terms of computational difficulty and solution character when these aspects are considered as a function of parameters such as the ratio of the number of constraints to the number of variables. According to Gent and Walsh (1994), the phase transition between satisfiability (feasibility) and unsatisfiability (unfeasibility) for the discrete (optimisation) problem appears as α (the ratio of its constraints to its variables) is varied.[1]

---

[1] The experimental work on this $\mathcal{NP}$-complete problem was done using the random k-SAT model, as it has several features that make it useful for benchmarking. SAT (propositional satisfiability) is the problem of deciding whether there is an assignment for a variable in a propositional formula that makes the formula true (Gent and Walsh, 1994).

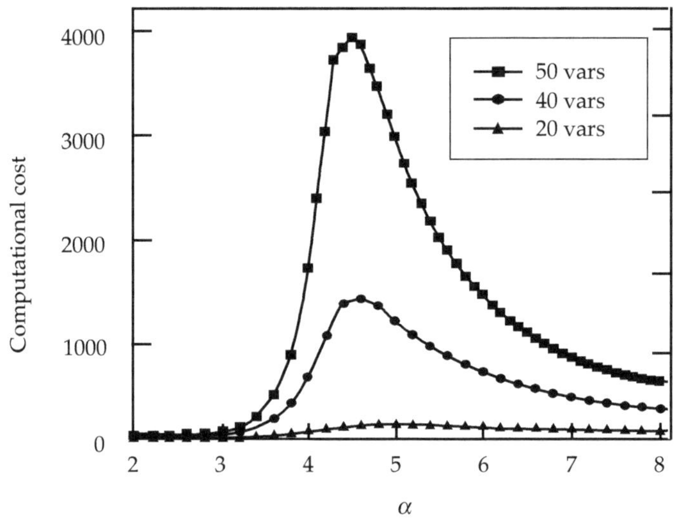

Figure 6.2 Solution complexity patterns for a typical $\mathcal{NP}$-complete problem with different numbers of variables

Figure 6.2 shows a typical *easy/hard/less-hard* pattern in computation cost (difficulty) for a given MIP problem as a function of α. Note that there are relatively few constraints and many variables – which mean that the problem is relatively *easy* to solve since it is underconstrained – at low values of α. On the other hand, at high values of α, the problem is overconstrained and is almost always unfeasible (*less-hard* region). As well as these two regions, a third region corresponding to the edge between the aforementioned regions can be discerned; this is where the problems are hardest to solve (*hard* region) (see Etherington and Parkes, 2004).

Depending on the problem structure (*i.e.*, the number of constraints and variables of an MIP problem for a given sewer network), the value of α can be located anywhere along the x-axis of Figure 6.2. Changing the level of complexity of the MIP problem may not be achieved by suppressing a small number of logical variables or adding some additional constraints. Therefore, this approach needs to be applied by considering a strong reduction in switching dynamics, achieved by either suppressing many logical variables or adding a large number of constraints in order to change the value of α considerably. This procedure reduces computation time by simply reducing Γ and in turn reducing the number of feasible system modes. Experimental results have explicitly shown the clear relation between the runtime of an $\mathcal{NP}$-complete problem and α (Nudelman *et al.*, 2004). Phase-transition behaviour was studied in depth by Kalapala (2005), and has been applied for instance in multi-vehicle task assignment problems – see Earl and D'Andrea (2005) – among other applications.

6.2.2 Strategies to Deal with the Complexity of HMPC

Given the computation time issue, it is necessary to explore some ways to relax and/or simplify the discrete optimisation problem and to find methodologies that can make HMPC designs practically applicable to large problems, such as the RTC of sewer networks. Most of the hybrid control approaches presented in the literature were applied to rather small examples. For large-scale systems, there is currently no standard strategy to relax the problem in order to find a trade-off between optimality and an acceptable amount of computation time.

Control strategies have been proposed where the HMPC problem is relaxed to make it computationally tractable. In Borrelli *et al.* (2005b), a *decentralised control* approach to HMPC was presented. The class of systems considered in that work consisted of dynamically uncoupled subsystems, but global control objectives were formulated with a global cost function.

A number of authors have also presented methods where the objective is to reduce complexity off-line. In Borrelli *et al.* (2005a), an explicit solution to the constrained finite-time optimal control problem was presented for discrete-time linear hybrid systems. *Mode enumeration* (ME), as reported in Geyer *et al.* (2003), is an off-line technique to compute and explicitly enumerate the feasible modes of PWA models. This technique allows the designer to understand the real complexity of the system and moreover to take advantage of its topology. Thus, once the feasible modes have been identified, the model can be efficiently translated to a specific hybrid system framework such as MLD, MMPS, LC, *etc.* (Heemels *et al.*, 2001). The difference between this technique and the similar problem solved in Bemporad (2004) is the computation of cells in the hyperplane formed by the input-state space, thus yielding the PWA model. The approximation given in Bemporad (2004) is based on multi-parametric programming and MIP, and it deals directly with the MLD form of the system model.

In order to apply MPC to a closed-loop control scheme, ME allows unnecessary modes of the resulting system to be pruned, and it reduces combinatorial explosion for algorithms. This allows modes to be cut from the complete prediction model over a given horizon. Even though the technique has been reported to be very efficient (Geyer *et al.*, 2005), its application to large-scale systems implies huge computational complexity due to the densely and explicitly partitioned state space. This off-line procedure can determine a few regions to prune, and the result can remain a hybrid model with many logical variables, which implies a large-scale MIP problem.

In this chapter, an HMPC strategy that limits the number of feasible nodes in the MIP problem on-line is proposed. This approach is performed by adding constraints to the MIP problem based on insight into the system dynamics. The idea is to help the MIP solver by making cuts to the search space. In this way, the main source of complexity, namely the combinatorial explosion related to the binary search tree, is reduced at the expense of a suboptimal solution. It has been recognised in the MPC literature that even though a solution is only suboptimal, its stability can often be proven (Mayne *et al.*, 2000), so the stability of this solution is then proven using reported results for HMPC (Lazar *et al.*, 2006). Unfeasibility is avoided by restricting the number of possible combinations of the model's Boolean variables in order to obtain sequences close to the nominal feasible sequence.

6.3 MPC of Hybrid Systems Incorporating Mode Sequence Constraints

This section details the proposed suboptimal approach, which involves limiting the commutation of the system between its dynamical models. This limitation is performed by considering a given mode sequence reference. The section also presents the conditions needed for feasibility and closed-loop stability within the framework of the proposed suboptimal approach. Some definitions and results of this section closely adhere to Sections II and III in Lazar *et al.* (2006). Other approaches to obtaining suboptimal solutions and stability guarantees for MPC schemes applied to hybrid systems are reported in Bertsekas (2005), Habibi *et al.* (2006), Lazar *et al.* (2008) and Lazar and Heemels (2009).

6.3.1 Description of the Approach

Assume that there are no disturbances and that the polyhedra $\mathbb{X}$ and $\mathbb{U}$, which both contain the origin, represent the state and input constraints, respectively. The state x_k and the control signal u_k as defined by the MLD (2.8) are mapped as shown in (2.1):

$$x_{k+1} = g(x_k, u_k), \tag{6.3}$$

where g is a discontinuous function for MLD forms. It is assumed that the origin is an equilibrium state with $u_k = 0$, in other words $g(0,0) = 0$.

Consider the sequences (2.2), (2.4), (2.12) and (2.13) presented and explained in Section 2.2.1. In what follows, the sequence $\Delta_k(x_k, \mathbf{u}_k)$ in (2.12) will be denoted the *mode sequence*. Let $\mathbb{X}_T$ (the target state set) contain the origin. Also let

$$\bar{\Delta}_k = \left(\bar{\delta}_{0|k}, \ldots, \bar{\delta}_{H_p-1|k}\right) \in \{0,1\}^{r_\ell \times H_p} \tag{6.4}$$

be a *reference sequence* of binary variables $\bar{\delta}_k$ of the same dimension as Δ_k. The mode sequence constraints are now defined by the following inequalities:

$$\sum_{k=0}^{H_p-1} |\bar{\delta}_k^i - \delta_k^i| \leq M_i, \tag{6.5a}$$

$$\sum_{i=1}^{r_\ell} \sum_{k=0}^{H_p-1} |\bar{\delta}_k^i - \delta_k^i| \leq M, \tag{6.5b}$$

for $i = 1 \dots r_\ell$, where $M, M_i \in \mathbb{Z}_+$ are given bounds on the number of switches from the reference sequence $\bar{\Delta}_k$. These constraints define the related sets $\mathscr{D}_{M_i}(\bar{\Delta}_k) \subseteq \{0,1\}^{r_\ell \times H_p}$ and $\mathscr{D}_M(\bar{\Delta}_k) \subseteq \{0,1\}^{r_\ell \times H_p}$ in the following manner:

$$\mathscr{D}_{M_i}(\bar{\Delta}_k) = \left\{ \Delta_k \in \{0,1\}^{r_\ell \times H_p} | \sum_{k=0}^{H_p-1} |\bar{\delta}_k^i - \delta_k^i| \leq M_i \right\}, \tag{6.6a}$$

$$\mathscr{D}_M(\bar{\Delta}_k) = \left\{ \Delta_k \in \{0,1\}^{r_\ell \times H_p} | \sum_{i=1}^{r_\ell} \sum_{k=0}^{H_p-1} |\bar{\delta}_k^i - \delta_k^i| \leq M \right\}, \tag{6.6b}$$

for $i = 1 \dots r_\ell$. The dependence of Δ_k on x_k and $\mathbf{u}_k$ is omitted for compactness. These sets contain the sequences Δ_k, which have a limited number of differences from a reference sequence $\bar{\Delta}_k$. If $\Delta_k, \delta_k, \bar{\Delta}_k$ and $\bar{\delta}_k$ are considered to be binary strings, the inequalities that define these sets limit the Hamming distance[2] between the strings involved. In what follows, the discussion will be limited to the set $\mathscr{D}_M(\bar{\Delta}_k)$ for the sake of compactness. The proof of stability is exactly the same if M is replaced with M_i.

Simple combinatorial counting of the sizes of sets (6.6) (*i.e.*, their *cardinality*) defined by inequalities (6.5) gives the following results:

$$N_{M_i} = \left(\sum_{j=1}^{M_i} \frac{N!}{j!(N-j)!} \right)^{r_\ell}, \tag{6.7}$$

$$N_M = \sum_{j=1}^{M} \frac{Nr_\ell!}{j!(Nr_\ell - j)!}. \tag{6.8}$$

The number N_M is the number of different Δ_k sequences that fulfil (6.5b), while N_{M_i} is the number of different Δ_k sequences related to (6.5a).

The class of admissible input sequences (2.3) is now defined with respect to $\mathbb{X}_T$ and set $\mathscr{D}_M$ as

$$\mathscr{U}_{H_p}(x_k, \bar{\Delta}_k) \triangleq \left\{ \mathbf{u}_k \in \mathbb{U}^{H_p} | \mathbf{x}_k(x_k, \mathbf{u}_k) \in \mathbb{X}^{H_p}, x_{H_p|k} \in \mathbb{X}_T, \Delta_k(x_k, \mathbf{u}_k) \in \mathscr{D}_M(\bar{\Delta}_k) \right\}. \tag{6.9}$$

[2] In information theory, the Hamming distance between two strings of equal length is the number of times that the corresponding positions in the strings have different symbols. Put another way, it measures the number of substitutions required to change one into the other, or the number of errors that transformed one string into the other.

Remark 6.1. Note that this set can be characterised exactly by a mixed-integer linear inequality (2.8c). ◇

The MPC problem described in Section 2.2.1 is now stated in a similar way to that in Lazar *et al.* (2006). Some sequences and definitions given in that work are rewritten here for completeness.

Problem 6.1 (MPC Problem with Mode Sequence Constraints). Assume that the target set $\mathbb{X}_T \subset \mathbb{X}$ and $H_p \in \mathbb{Z}_{\geq 1}$ are given. Minimise the cost function

$$J(x_k, \mathbf{u}_k) = F(x_{H_p|k}) + \sum_{i=0}^{H_p-1} L(x_{i|k}, u_{i|k}) \tag{6.10}$$

over $\mathbf{u}_k \in \mathscr{U}_{H_p}(x_k, \bar{\Delta}_k)$, where $F : \mathbb{R}^n \to \mathbb{R}_+$ and $L : \mathbb{R}^n \times \mathbb{R}^m \to \mathbb{R}_+$ are functions that fulfil $F(0) = 0$ and $L(0,0) = 0$.

An initial state $x_0 \in \mathbb{X}$ is feasible if there is a reference sequence $\bar{\Delta}_k$ such that $\mathscr{U}_{H_p}(x_0, \bar{\Delta}_k) \neq \emptyset$. Hence, Problem 6.1 is feasible if there is a feasible $x \in \mathbb{X}$ such that $\mathscr{U}_{H_p}(x, \bar{\Delta}_k) \neq \emptyset$. Let the set $\mathbb{X}_f(H_p) \subseteq \mathbb{X}$ denote the *set of feasible states*. The function

$$V_{\text{MPC}}(x_k) = \min_{\mathbf{u}_k \in \mathscr{U}_{H_p}(x_k, \bar{\Delta}_k)} J(x_k, \mathbf{u}_k), \tag{6.11}$$

related to Problem 6.1, is called the *MPC value function*. It is assumed that there is an optimal control sequence (2.6)

$$\mathbf{u}_k^* = \left(u_{0|k}^*, u_{1|k-1}^*, \ldots, u_{H_p-1|k}^*\right)$$

for the above problem and that any state $x_k \in \mathbb{X}_f(H_p)$. Using the receding horizon philosophy, the MPC control law is defined as in (2.7):

$$u_{\text{MPC}}(x_k) \triangleq u_{0|k}^*, \tag{6.12}$$

where $u_{0|k}^*$ is the first element of $\mathbf{u}_k^*$.

Remark 6.2. The reference sequence between samples k and $k+1$ can be chosen appropriately by considering the sequence $\Delta_k^* = (\delta_{0|k}^*, \delta_{1|k}^*, \ldots, \delta_{H_p-1|k}^*)$ obtained from the solution of Problem 6.1 at time k. Given a $\vartheta \in \mathbb{U}$ that fulfils $x_{H_p+1|k} = g(x_{H_p|k}^*, \vartheta) \in \mathbb{X}_T$, the reference mode sequence at time $k+1$ is set to

$$\bar{\Delta}_{k+1} = \left(\delta_{1|k}^*, \ldots, \delta_{H_p-1|k}^*, \delta_+(x_{H_p|k}^*, \vartheta)\right), \tag{6.13}$$

where $\delta_+(x_{H_p|k}^*, \vartheta)$ is found using the system equations in (6.3). ◇

However, to determine the reference sequence (6.13), the input ϑ must fulfil certain specific conditions (Mayne *et al.*, 2000). In this sense, and according to Lazar *et al.* (2006), both feasibility and stability can be ensured by applying a terminal cost and constraint set method such as that in Mayne *et al.* (2000), but with the conditions and assumptions adapted to hybrid systems. Therefore, the following assumption is now presented in order to prove the stability of the closed-loop system (6.3) and (6.12). This assumption is presented unchanged from Lazar *et al.* (2006).

Assumption 6.1 (Lazar *et al.* (2006)) *Assume that there are strictly increasing, continuous functions $\alpha_1, \alpha_2 : \mathbb{R}_+ \to \mathbb{R}_+$ that fulfil $\alpha_1(0) = \alpha_2(0) = 0$, a neighbourhood of the origin $\mathcal{N} \subset \mathbb{X}_f(H_p)$, and a nonlinear, possibly discontinuous function $h : \mathbb{R}^n \to \mathbb{R}^m$ such that $\mathbb{X}_T \subset \mathbb{X}_{\mathbb{U}}$ with $0 \in \operatorname{int}(\mathbb{X}_T)$ is a positively invariant set for system (6.3) in a closed loop with $u_k = h(x_k)$. $\mathbb{X}_{\mathbb{U}}$ denotes the safe set with respect to state and input constraints for $h(\cdot)$. Furthermore,*

$$L(x,u) \le \alpha_1(\|x\|), \quad \forall x \in \mathbb{X}_f(H_p), \forall u \in \mathbb{U}, \tag{6.14a}$$

$$F(x) \ge \alpha_2(\|x\|), \quad \forall x \in \mathcal{N} \quad \text{and} \tag{6.14b}$$

$$F(g(x,h(x))) - F(x) + L(x,h(x)) \le 0, \quad \forall x \in \mathbb{X}_T. \tag{6.14c}$$

♦

A theorem for the stability of MPC controllers with mode sequence constraints is now presented. Its proof closely follows the proof presented in Lazar *et al.* (2006), but for completeness it is repeated here, considering the concepts and sequences defined for the proposed approach. The proof rests on Lyapunov stability results for systems with discontinuous dynamics developed in Lazar (2006), but it also takes into account mode sequence constraints.

Theorem 6.1. *For fixed H_p, suppose that Assumption 6.1 holds. Then:*

1. *It also holds that if Problem 6.1 is feasible at time k for state $x_k \in \mathbb{X}$, then Problem 6.1 is feasible at time $k+1$ for state $x_{k+1} = g(x_k, u_{\text{MPC}}(x_k))$ and $\mathbb{X}_T \subseteq \mathbb{X}_f(H_p)$*
2. *It holds that $\mathbb{X}_T \subset \mathbb{X}_f(H_p)$*
3. *The origin of the MPC closed-loop system formed by applying control law (6.12) to plant (6.3) is asymptotically stable in the Lyapunov sense for the initial conditions in $\mathbb{X}_f(H_p)$.*

Proof. Consider the shifted sequence

$$\mathbf{u}^s_{k+1} \triangleq \left(u^*_{1|k}, \dots, u^*_{H_p-1|k}, h(x_{H_p-1|k+1})\right), \tag{6.15}$$

where $x_{H_p-1|k+1}$ is the state at prediction time $H_p - 1$, obtained at time $k+1$ by applying the input sequence $u^*_{1|k}, \dots, u^*_{H_p-1|k}$ to system (6.3) with the initial condition $x_{0|k+1} \triangleq x^*_{1|k} = x_{k+1} = g(x_k, u_{\text{MPC}}(x_k))$. Note that $x_{H_p-1|k+1} = x^*_{H_p|k}$.

1. If Problem 6.1 is feasible at time $k \in \mathbb{Z}_+$ for state $x_k \in \mathbb{X}$, then there is a mode sequence reference $\underline{\Delta}_k$ such that $\mathscr{U}(x_k, \underline{\Delta}_k)$ is not empty. The optimal solution to Problem 6.1 is denoted $\mathbf{u}_k^*$. It then follows that $x_{H_p-1|k+1} \in \mathbb{X}_T$. Due to Remark 6.2 and the positive invariance of $\mathbb{X}_T \in \mathbb{X}_{\mathbb{U}}$, it holds that $x_{H_p|k+1} \in \mathbb{X}_T$ and $\mathbf{u}_{k+1}^s \in \mathscr{U}(x_{k+1}, \underline{\Delta}_{k+1})$. This implies that Problem 6.1 is feasible at time $k+1$ for x_{k+1} and mode sequence reference $\underline{\Delta}_{k+1}$.
2. Let $\tilde{\mathbf{x}}(x_k) = (\tilde{x}_{1|k}, \tilde{x}_{2|k} \ldots \tilde{x}_{H_p|k})$ denote the state sequence generated by the system $x_{k+1} = g(x_k, h(x_k))$ from initial state $\tilde{x}_{0|k} \triangleq x_k \in \mathbb{X}_T$. Let $\tilde{\mathbf{u}}_k$ denote the corresponding control signal. Since $\tilde{\mathbf{x}}_k \in \mathbb{X}_T^{H_p}$, then $\tilde{\mathbf{u}}_k \in \mathbb{U}$, according to Assumption 6.1. A candidate reference sequence $\underline{\Delta}_k$ such that $\mathscr{U}_k(x_k, \underline{\Delta}_k) \neq \emptyset$ is the one related to $\tilde{\mathbf{u}}_k$ and $\tilde{\mathbf{x}}_k$.
3. Consider again the state sequence $\tilde{\mathbf{x}}_k(x_k)$. Since $\tilde{\mathbf{x}}_k(x_k) \in X_T^{H_p}$, inequality (6.14c) from Remark 6.2 holds for all elements in $\tilde{\mathbf{x}}_k$, yielding

$$\begin{aligned}
F(\tilde{x}_{1|k}) - F(\tilde{x}_{0|k}) + L(\tilde{x}_{0|k}, h(\tilde{x}_{0|k})) &\leq 0\\
F(\tilde{x}_{2|k}) - F(\tilde{x}_{1|k}) + L(\tilde{x}_{1|k}, h(\tilde{x}_{1|k})) &\leq 0\\
&\vdots\\
F(\tilde{x}_{H_p|k}) - F(\tilde{x}_{H_p-1|k}) + L(\tilde{x}_{H_p-1|k}, h(\tilde{x}_{H_p-1|k})) &\leq 0.
\end{aligned}$$

From these inequalities, by optimality and considering Remark 6.2, it follows that

$$V_{\mathrm{MPC}}(x_k) \leq J(x_k, \tilde{\mathbf{u}}_k) \leq F(x_k) \leq \alpha_2(\|x_k\|), \quad \forall x_k \in \widetilde{\mathscr{N}},$$

where $\widetilde{\mathscr{N}} = \mathscr{N} \cap \mathbb{X}_T$. Again, using optimality,

$$\begin{aligned}
V_{\mathrm{MPC}}(x_{k+1}) - V_{\mathrm{MPC}}(x_k) &= J(x_{k+1}, \mathbf{u}_{k+1}^*) - J(x_k, \mathbf{u}_k^*)\\
&\leq J(x_{k+1}, \mathbf{u}_{k+1}^1) - J(x_k, \mathbf{u}_k^*)\\
&= -L(x_k, u_{\mathrm{MPC}}(x_k)) + F(x_{H_p|k+1}) - F(x_{H_p|k}^*)\\
&\quad + L(x_{H_p|k}^*, h(x_{H_p|k}^*)).
\end{aligned}$$

Then, as $x_{H_p|k}^* \in \mathbb{X}_T$, and because of condition (6.14c) in Assumption 6.1, it holds that

$$V_{\mathrm{MPC}}(g(x_k, u_{\mathrm{MPC}}(x_k))) - V_{\mathrm{MPC}}(x_k) \leq -L(x_k, u_{\mathrm{MPC}}(x_k)) \leq -\alpha_1(\|x_k\|), \\ \forall x_k \in \mathbb{X}_f(H_p).$$

Since $\mathbb{X}$ is compact and $\mathbb{X}_f \subset \mathbb{X}$, then according to item 1., $\mathbb{X}_f(H_p)$ is positively invariant. Let x_k be a state reached with the closed-loop system (6.3) and (6.12) from initial state x_0. Choose any $\eta > 0$ such that the ball $\mathscr{B}_\eta \triangleq \{x \in \mathbb{R}^n \,|\, \|x\| \leq \eta\}$ satisfies $\mathscr{B}_\eta \subset \widetilde{\mathscr{N}}$. It is possible to choose any $0 < \varepsilon \leq \eta$ and $\sigma \in (0, \varepsilon)$ such that $\alpha(\sigma) < \alpha(\varepsilon)$. For any $x_0 \in \mathscr{B}_\sigma \subset \mathbb{X}_f(H_p)$, due to the positive invariance of $\mathbb{X}_f(H_p)$, it follows that

$$\ldots \leq V_{\text{MPC}}(x_{k+1}) \leq V_{\text{MPC}}(x_{k+1}) \leq \cdots \leq V_{\text{MPC}}(x_0) \leq \alpha_2(\|x_0\|) \leq \alpha_2(\sigma) \leq \alpha_1(\varepsilon).$$

Since $V_{\text{MPC}}(x) \geq \alpha_1(\varepsilon)$ for all $x \in \mathbb{X}_f(H_p)\backslash\mathscr{B}_\varepsilon$, it follows that $x_k \in \mathscr{B}_\varepsilon$ for all $k \in \mathbb{Z}_+$. □

Therefore, according to Assumption 6.1 and the results given by Theorem 6.1, $\delta_+(x^*_{H_p|k}, \vartheta)$ in Remark 6.2 can be computed using the local control law $\vartheta = h(x_k)$, ensuring feasibility and stability. Assuming cost function (6.10) defined by

$$F(x_k) = \|Px_k\|_p \qquad \text{and} \qquad L(x_k, u_k) = \|Qx_k\|_p + \|Ru_k\|_p,$$

and the local control law set as $h(x_k) = Kx_k$, the computation of the weight matrix P and the piecewise linear state-feedback gain K that fulfils Assumption 6.1 and Remark 6.2 are reported in Lazar (2006) for the norms $p = 1, \infty$ and $p = 2$. These computations are done off-line using different methods and algorithms discussed in the mentioned reference and the references therein.

6.3.2 *Practical Issues*

One important practical problem with the proposed method is finding $\bar{\Delta}_k$ such that $\mathscr{U}_{H_p}(x_k, \bar{\Delta}_k)$ is not empty and the MIP problem with mode sequence constraints has a solution. When states are measured and disturbances are present, the assumption that $x_{0|k} = x_{1|k-1}$ will not hold and the shifted sequence from the previous sample will not necessary be feasible. Finding $\bar{\Delta}_k$ using (6.13) is therefore not an option.

The issue of finding $\bar{\Delta}_k$ for distinct cases of state and input constraints will now be analysed. The main method that is used to find this sequence is to solve a CSP.[3] A natural candidate solution that may be close to the optimum is the shifted sequence from the last sample, given by (see the proof of Theorem 6.1)

$$\mathbf{u}^s_{k+1} \triangleq \left(u^*_{1|k}, \ldots, u^*_{H_p-1|k}, h(x_{H_p-1|k+1})\right). \tag{6.16}$$

This control sequence can be used to simulate the open-loop system. If all constraints are respected, $\mathbf{u}^s_k$ is a feasible solution. If the measured state is close to the predicted state, it is reasonable to believe that this sequence will at least provide a good initial guess that is close to the optimum.

---

[3] Depending on the case, the CSP is equivalent to a simulation of the open-loop model. However, when input, state and/or output constraints are present, only a CSP statement makes sense. CSP concepts are presented and discussed in Chapter 8.

No State Constraints

If $\mathbb{X} = \mathbb{R}^n$ (no state constraints) and the system (6.3) is stable, then using $\mathbf{u}_k^s$ defined in (6.16) in an open-loop simulation from the new initial state $x_{0|k}$ results in a sequence $\underline{\Delta}_k$ that can be used to form $\mathscr{U}_{H_p}(x_k, \underline{\Delta}_k)$. If $\mathbf{u}_k^s$ is not available or system (6.3) is unstable, the way to find the sequence $\underline{\Delta}_k$ is unclear, and it is necessary to use heuristic knowledge of the system. As a possible alternative, a feasible input sequence can be set using the local control law

$$u_k = \mathrm{sat}(h(x_k)), \quad x_k \in \mathbb{X}_f(H_p), \tag{6.17}$$

for $k = 0, \ldots, H_p - 1$, and the system is again simulated in an open loop from the initial state $x_{0|k}$. Function $u_k = \mathrm{sat}(\cdot)$ guarantees that $u_k \in \mathbb{U}$.

State Constraints

State constraints are generally related to either physical constraints of the model, such as conservation equations and physical limitations of the process, or to control objectives.

If the open-loop simulation (*i.e.*, a CSP) fails and some constraints are violated, in the worst case, $\underline{\Delta}_k$ is found by searching for a feasible trajectory for the problem without mode sequence constraints from the new initial state. This is, in turn, an MIP feasibility problem. The reduction in time that can be achieved with the presented methodology then depends on the complexity of feasibility problem compared to the optimisation problem – something that is difficult to analyse *a priori*. This is a restriction on the presented method, but if constraints related to safety or high risk are present in $\mathbb{X}$, and feasibility cannot be assured within a prespecified timeframe, neither the presented method nor any other HMPC strategy that depends on an MIP problem to find a feasible solution would be applicable in practice.

Constraint Management

Constraint management is an important issue in constrained predictive control (see Maciejowski, 2002). A common approach to dealing with unfeasibilities is to change constraints from *hard* to *soft*; that is, to add terms containing slack variables of the constraints to the cost function. If the changed constraints represent physical characteristics, the resulting control signal may be of little use, as the model from which the control signal is obtained may not fulfil basic physical laws. If the constraints are related to safety considerations, the resulting control signal may not be applicable either.

Constraint management is equally important in the presented scheme, as a straightforward way of obtaining an initial feasible solution is to change any unfulfilled constraints in $\mathbb{X}$, when $\mathbf{u}_k^s$ is used in open-loop simulation, into soft constraints. As mentioned previously, this approach is only appropriate when the relaxed constraints do not represent physical or safety characteristics of the system.

When forming the cost function containing the slack variables relating to the soft constraints, it is often the case that some constraints have higher priority than others. The common way to deal with distinct priorities is to assign weights to the slack variables that reflect their importance. These weights are generally selected through trial and error procedures involving simulations of typical disturbance and reference value scenarios. If the relative importance of the relaxed constraints is known, objective prioritisation schemes implemented with propositional logic (Tyler and Morari, 1999) represent an interesting option, as these schemes are implemented with MIP solvers.

Finding a Feasible Solution with Physical Knowledge and Heuristics

The physics or heuristic knowledge of the system can often be used to find a feasible solution that fulfils the physical constraints of the system. For instance, all integer variables have fixed values in the steady state, a fact that could be used when determining the sequence Δ_k.

State constraints representing physical limitations can often be incorporated into the hybrid model using propositional logic. As an example, consider a tank that has an upper limit on its level and its inflow controlled with a valve. The upper limit on the tank could be modelled by adding a constraint to the optimisation problem so that any controlled signal to the valve that causes the level to surpass the physical limit would be unfeasible in the optimisation problem. Within the hybrid modelling framework, a logical statement could be incorporated that guarantees that the inflow to the tank would never cause the level to surpass its physical limit, irrespective of the control signal to the valve.

This hybrid modelling approach actually represents the physical behaviour better and enables the removal of a state constraint that would allow unfeasibility to occur during the open-loop simulation. On the other hand, it would increase the amount of Boolean variables in the system hybrid model.

Suboptimal Approach and Disturbances

Now consider system (6.3), including disturbances. It is rewritten as follows:

$$x_{k+1} = g(x_k, u_k, d_k), \tag{6.18}$$

where $d_k \in \mathbb{R}_+^{md}$ denotes the vector of bounded disturbances. In the presence of uncertainty and disturbances, a reference sequence $\overline{\Delta}_k$ cannot be obtained in the manner proposed in Section 6.3, as the measured state at time $k \in \mathbb{Z}_{\geq 2}$ will not correspond to that predicted from the previous sample ($x_{0|k} \neq x_{1|k-1}$). This means that the sequence $\overline{\delta}_k$ will not necessarily be feasible, and this problem will appear at each time instant. In the worst case, this issue can be reduced to finding a feasible solution to an MIP problem. This is also the case when measured disturbances or reference signals are taken into account over the prediction horizon. These can be transformed into an equivalent MIP feasibility problem given an extended initial state.

6.4 Suboptimal HMPC Strategy for Sewer Networks

Having presented and discussed the proposed suboptimal strategy, it can now be applied in the HMPC design for the sewer network. Indeed, this strategy was inspired by large-scale systems such as the type considered in this book, since computation time is an important issue when implementing the control of such systems on-line. Since rain is the main influence considered in sewer network control design, this section will use the ideas presented in Section 6.3.2 to explain how the suboptimal strategy is applied. The main results are given below and the corresponding conclusions are outlined (note that the simulation results reported in Chapter 5 were obtained using the suboptimal controller).

6.4.1 Suboptimal Strategy Set-up

First of all, two facts about the case study are considered:

1. Sewer networks are generally designed to be stable systems according to the modelling framework and the hierarchical control philosophy (Papageorgiou, 1985). Computed control signals only modify the values of performance indices. For this reason, a target state set $\mathbb{X}_T$ is not considered.
2. The associated optimisation always has a feasible solution, as the state constraints can be relaxed (soft constraints). Only the states related to real tanks within the virtual tank modelling methodology are hard constraints, but these restrictions can be passed on to their related inflows (manipulated flow at the control gate upstream).

Let

$$\mathbf{d}_k = \left(d_{0|k}, d_{1|k}, \ldots, d_{H_p-1|k}\right) \in \mathbb{R}_+^{md} \tag{6.19}$$

be the sequence containing the measured disturbances. Generally, only $d_{0|k}$ is measured while the other values are predicted over the prediction horizon (see Section 5.1.5).

To obtain a reference mode sequence $\overline{\Delta}_{k+1}$ from the optimal sequence Δ_k^*, the optimal state value $v_{H_p|k}^* \in \mathbf{v}_k$, the value $d_{H_p-1|k} \in \mathbf{d}_k$ and the last value of the sequence

$$\mathbf{q_u}_k^* \triangleq \left(q_{u0|k}^*, q_{u1|k}^*, \ldots, q_{uH_p-1|k}^*\right) \tag{6.20}$$

are used to compute the last value of $\overline{\Delta}_{k+1}$, $\delta_+(v_{H_p|k}^*, \vartheta, d_{H_p-1|k})$. However, it could be that $\vartheta = q_{uH_p-1|k}^*$ makes the optimisation problem unfeasible, as the value of $q_{uH_p-1|k}^*$ can be such that the physical constraints of the system in (5.52) are not respected. Hence, considering the discussion in Section 6.3.2, and using an MLD$_P$ (as explained in Section 5.3.2), $q_{uH_p-1|k}^*$ is validated by simulating the system with the mode-detailed MLD$_P$ model. A validated control signal $\tilde{q}_u$ is then obtained which is used to set $\delta_+(v_{H_p|k}^*, \tilde{q}_u, d_{H_p-1|k})$ within the reference sequence for $k+1$.

Remark 6.3. The computation of $\tilde{q}_u$ from $q_{uH_p-1|k}^*$ using the system model MLD$_P$ can be seen as a CSP (Jaulin *et al.*, 2001a). This technique is applied to refine a system set using its constraints. ◊

Finally, the control strategy applied is as follows. At time k,

1. Obtain the reference sequence $\overline{\Delta}_k$ using Δ_{k-1}^* and the MLD$_P$ model
2. Add the corresponding set of constraints in (6.5) to the MIP problem behind the HMPC controller design
3. Solve the MIP problem and obtain a new sequence $\mathbf{q_u}_k^*$
4. Apply the control law (5.61) to the process.

Note that adding restrictions of type (6.5) to the MIP problem will not cause unfeasibility, as the trivial solution $\delta_k^i = \overline{\delta}_k^i$ always fulfils the constraints of the problem.

6.4.2 *Simulation of Scenarios*

The suboptimal strategy was applied by simulating the closed-loop system for the rain episodes listed in Table 3.3. The structure of the closed-loop scheme was the same as that used for simulations in Chapter 5. The durations of the scenarios were determined by the duration of the rain peak and the system's reaction time to that rain, since the sample requiring maximum CPU time generally occurred after the rain peak. The efficiency of the approach was measured in terms of not only CPU time but also system suboptimality for different values of M and M_i in each case.

Tests were performed beforehand where CPLEX parameters were modified to suit the purpose of the suboptimal HMPC strategy (ILOG, 2003a). In this case, the parameter that sets the maximum time for a call to the solver, namely `CPX_PARAM_TILIM`, was modified from its default value of 10^{75} s to be smaller than Δt in order to fulfil the time requirements of the RTC problem. However, in some scenarios, this was not enough time to find at least a feasible solution. Therefore, another parameter that sets the balance between the feasibility and optimality of the solver solutions, namely `CPX_PARAM_MIPEMPHASIS`, was also modified in order to generate feasible solutions (that can be used as suboptimal solutions) in less time (the default value of this parameter applies an equal weight to both feasibility and optimality). This change reduces the CPU time for each sample, but the system performance was also reduced, so this option was ultimately ruled out.

6.4.3 Main Results

Simulations were performed with both types of constraints in (6.5), and results were obtained using the Hybrid Toolbox for MATLAB® (Bemporad, 2006) and ILOG CPLEX 9.1 as an MIP solver (ILOG, 2003b). The CPLEX parameters discussed before were set to their default values.

The suboptimality level of the strategy was measured by comparing the value of the cost function for the system with the suboptimal controller $J_k^{\mathrm{s}}(\mathbf{q_u}_k, \Delta_k, \bar{\Delta}_k, \mathbf{z}_k, \nu_k)$, with the value of the cost function for the closed-loop system without suboptimal approaches $J_k^{\mathrm{n}}(\mathbf{q_u}_k, \Delta_k, \mathbf{z}_k, \nu_k)$. This comparison was expressed as

$$\mathscr{S}_k = \frac{J_k^{\mathrm{s}}(\mathbf{q_u}_k, \Delta_k, \bar{\Delta}_k, \mathbf{z}_k, \nu_k)}{J_k^{\mathrm{n}}(\mathbf{q_u}_k, \Delta_k, \mathbf{z}_k, \nu_k)}. \tag{6.21}$$

A critical rain episode in the sense of CPU time was the one that occurred on October 17, 1999. Figure 6.3 (c) shows that the optimisation time in the unconstrained case rises by two orders of magnitude in one sample. At time 41 the optimisation time was 6.6 s, and at time 42 it was 689 s. As the sampling time was 300 s, this is unacceptable variability from an implementation perspective.

As mentioned before, the complexity of the optimisation problem can change due to a change in the number of feasible nodes. Feasible nodes in the MIP problem correspond to feasible sequences Δ_k. The reason why there are more feasible sequences at time 42 is demonstrated in Figure 6.3 (a) and (b). The rain peaks at time 42 for rain gauges P_{13} and P_{19}. Remember that the rain is predicted to be constant over the prediction horizon. This, in addition to the rising levels in the tanks, depending on the control sequence, means that many of the tanks could overflow at some point beyond H_p when the optimisation problem is formed at time 42. Tank overflow causes a Boolean

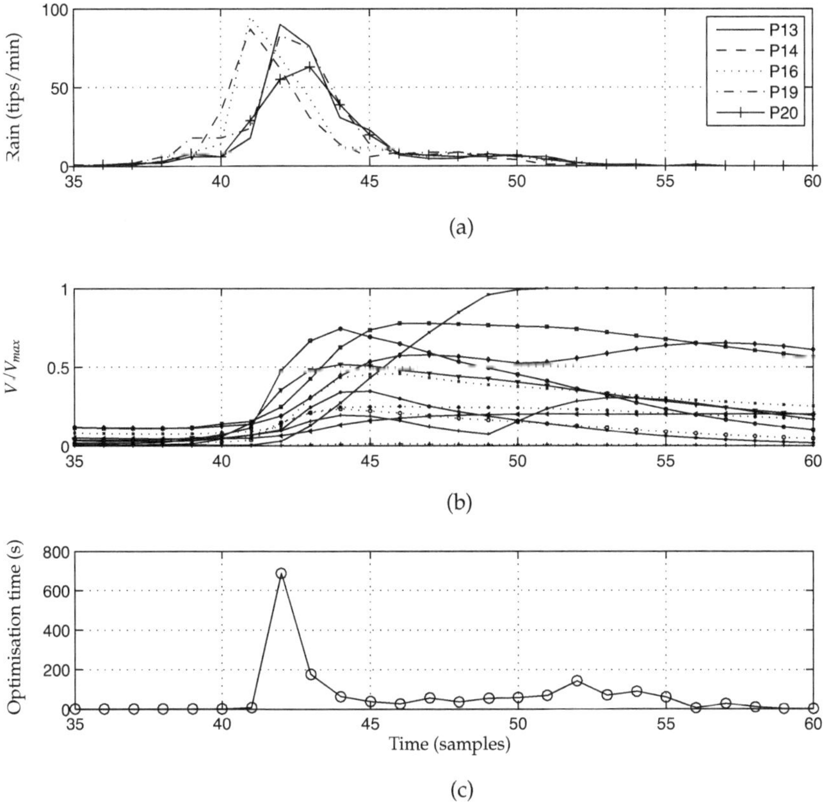

Figure 6.3 Rain episode 1999-10-17: (a) rain intensity, (b) virtual tank volumes normalised to their maximum volume, and (c) optimisation time

variable in the optimisation problem to change value, which, in turn, represents a change in the sequence Δ_k. The bottom line is that, at time 42, the combination of the rain prediction and the tank volumes lead to a big jump in the number of feasible Δ_k sequences. This, in turn, causes an increase in the number of feasible nodes. The number of nodes explored can be reported by the CPLEX solver. It was found that this number rises from 2520 at time 41 to 132,365 at time 42, which is an increase of several orders of magnitude.

Figure 6.4 shows the maximum optimisation time for the whole simulation scenario when M-type mode sequence constraints were implemented. In most cases, the maximum occurred at time 42, as expected. Figure 6.4 also shows that the optimisation time could be strongly reduced by implementing the mode sequence constraints. For $M \leq 4$, the maximum optimisation time does not surpass 50 s. As M increases the optimisation time rises until it settles around the optimisation time in the unconstrained case. In terms of nodes explored, compared to the number of feasible nodes as given in (6.8),

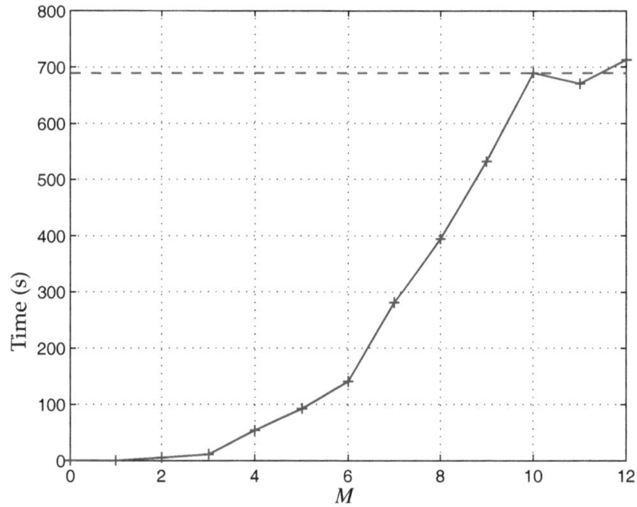

Figure 6.4 Maximum CPU times for the rain episode 99-10-17 when different values of M were used in (6.5b). *Dashed curve*, optimal simulation time

Table 6.2 Nodes explored as a function of M at sample 42

M	N_M	Nodes explored by CPLEX
1	133	66
2	8646	597

Table 6.2 demonstrates that the mode sequence constraints strongly affect the number of nodes explored by the MIP solver. Besides the fact that considerably fewer nodes were explored than in the unconstrained case quoted before, the number of nodes explored was always less than N_M.

According to Table 6.1, the rain episode that occurred on September 14, 1999 produced the greatest computational burden. Using the suboptimal approach, the computation time was reduced for small values of M or M_i with no significant reduction in system performance. In fact, the suboptimality was only 5% or so for the critical sample in the portion of the rain episode considered. Figure 6.5 shows the behaviour of $\mathscr{S}_k$ for each sample during the critical portion of the rain episode.

Regarding the reduction in CPU time for this rain episode, Figure 6.6 (a) plots M_i against the maximum CPU time for the closed-loop scheme with an HMPC controller that includes mode sequence constraints. Moreover, Figure 6.6 (b) shows the evolution of the CPU time for each sample and various values of M_i. It is apparent that decreasing M_i simplifies the MIP problem, meaning that the solver takes less time to find the suboptimal so-

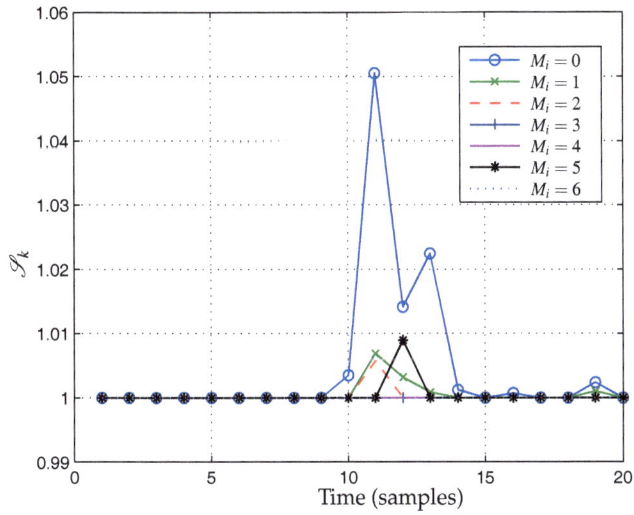

Figure 6.5 Suboptimality levels for the rain episode 99-09-14 when different values of M_i were used in (6.5a)

lution. When M_i (or M, depending on the case) is zero, the MIP problem is simply a QP or LP problem. Also bear in mind that when $M \geq H_p$, the mode sequence constraints do not have any influence on the optimality of the computed solution, as the sequences of Boolean variables can take any value. For this rain episode, in order to fulfil system time requirements, M_i should be strictly less than 2.

6.5 Suboptimal MPC Approach Based on Piecewise Linear Functions

In Chapter 5, a hybrid modelling approach based on MLD forms was used to model elements that exhibit switching behaviour in sewer networks. However, the inclusion of this switching behaviour in the MPC design increases the computation time of the control law. Therefore, some relaxation strategies (*e.g.*, in the modelling approach) are needed before these types of control schemes can be used for the RTC of large-scale sewer networks.

The aim of this section is to propose a control-oriented modelling approach that represents the sewage system using *piecewise linear functions*, following the ideas proposed by Schechter (1987). Henceforth, the terms *PWLF-based model* or simply *PWLF model* will be used to refer to this approach. The purpose of this modelling approach is to reduce the complexity

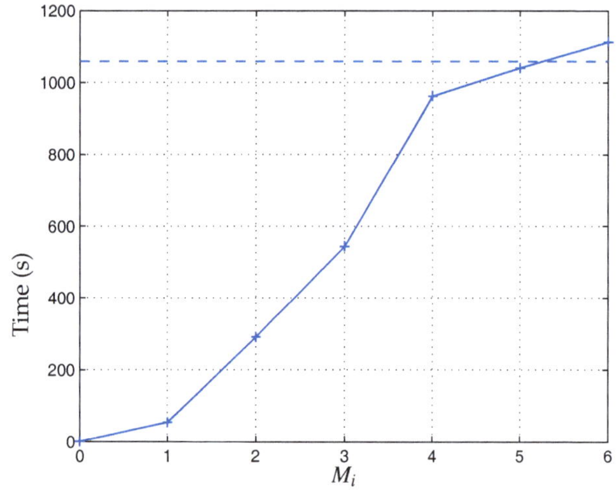

(a)

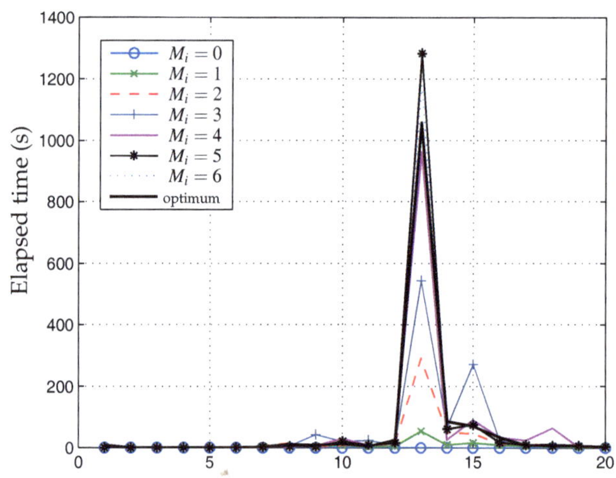

(b)

Figure 6.6 CPU times for different values of M_i in (6.5a) for the rain episode 99-09-14: (a) maximum CPU time for the scenario, and (b) evolution of the CPU time for each sample. In (a), the *dashed curve* shows the optimal simulation time

of the MPC problem by avoiding the logical variables introduced when the MLD form of the system is obtained for predictive purposes.

The idea behind the PWLF-based modelling approach is to get a description of the network using functions that, despite being discontinuous, are considered quasiconvex (Boyd and Vandenberghe, 2004). This can yield quasiconvex optimisation problems associated with the nonlinear MPC strategy employed in the RTC designs of sewage systems (Cembrano *et al.*, 2004). In this way, the resulting optimisation problems do not include Boolean variables, which saves computation time.

To show the benefits of the proposed PWLF approach, this section discusses its application to the design of MPC controllers for the RTC of sewer networks, and presents some results that were obtained with the BTC used as a case study.

6.5.1 PWLF Modelling Approach

The proposed alternative approach to hybrid modelling involves the use of continuous and monotonic functions to represent expressions that contain logical conditions which describe the behaviour of the weirs and the overflow capabilities of reservoirs. These phenomena are linked to the switching behaviour of the sewage system.

The properties of a monotonic and continuous function are very useful when optimisation-based control strategies are designed, as they allow a quasiconvex optimisation problem to be stated. According to Boyd and Vandenberghe (2004), the optimal global solutions to quasiconvex optimisation problems can be obtained using a bisection method, which is logarithmic with time. This is advantageous compared to the linear MIP problems that result when a pure hybrid approach based on an MLD or PWA approach is used. This type of model induces exponential complexity due to the handling of Boolean variables and the discrete optimisation required. The continuous and monotonic functions used for the modelling approach proposed here are as follows (see also Figure 6.7):

- Saturation function, defined as

$$\text{sat}(x,M) = \begin{cases} x & \text{if} \quad 0 \leq x \leq M, \\ M & \text{if} \quad x > M, \\ 0 & \text{if} \quad x < 0. \end{cases} \tag{6.22}$$

- Dead-zone function, defined as

$$\text{dzn}(x,M) = \begin{cases} x - M & \text{if} \quad x \geq M, \\ 0 & \text{if} \quad x < M. \end{cases} \tag{6.23}$$

Next, the constitutive elements of the sewer network described in Section 5.1 are expressed using this modelling approach. For convenience, Figure 6.8 shows the schemes for those elements. Note that the whole representation of a given sewer network modelled using this approach consists of

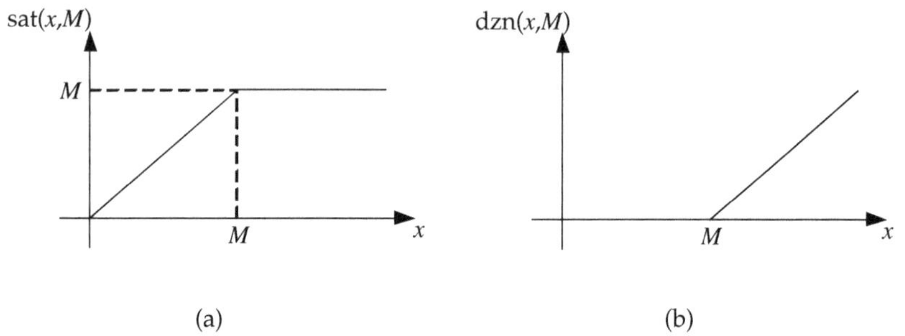

Figure 6.7 Continuous and monotonic piecewise functions used for sewer network modelling: (a) saturation function, and (b) dead-zone function

a set of equations rather than a matrix model such as that given by the hybrid modelling approach presented in Chapter 5. Actually, this kind of modelling approach was implemented to design an MPC law for the BTC years ago (Cembrano *et al.*, 2004). Software tools such as CORAL (Figueras *et al.*, 2002) can generate the set of equations that represent the behaviour and dynamics of the constitutive elements considered for a given sewer network topology. Once this set of equations has been obtained, CORAL computes a control sequence that minimises given performance indices. Using this strategy, suboptimal solutions of the control problem can be found according to a pre-established balance between complexity and computational burden.

Virtual Tanks

Using the PWLF approach, the tank outflows can be expressed as

$$q_{out\,k} = \beta\, \mathrm{sat}\left(v_k, \bar{v_k}\right), \tag{6.24}$$

$$q_{d\,k} = \frac{\mathrm{dzn}(v_k, \bar{v_k})}{\Delta t}, \tag{6.25}$$

which allows the difference equation for the tank volume to be written as

$$v_{k+1} = v_k + \Delta t (q_{in_k} - q_{d\,k} - q_{out\,k}), \tag{6.26}$$

where q_{in_k} considers all possible inflows, including the precipitation term (in flow units).

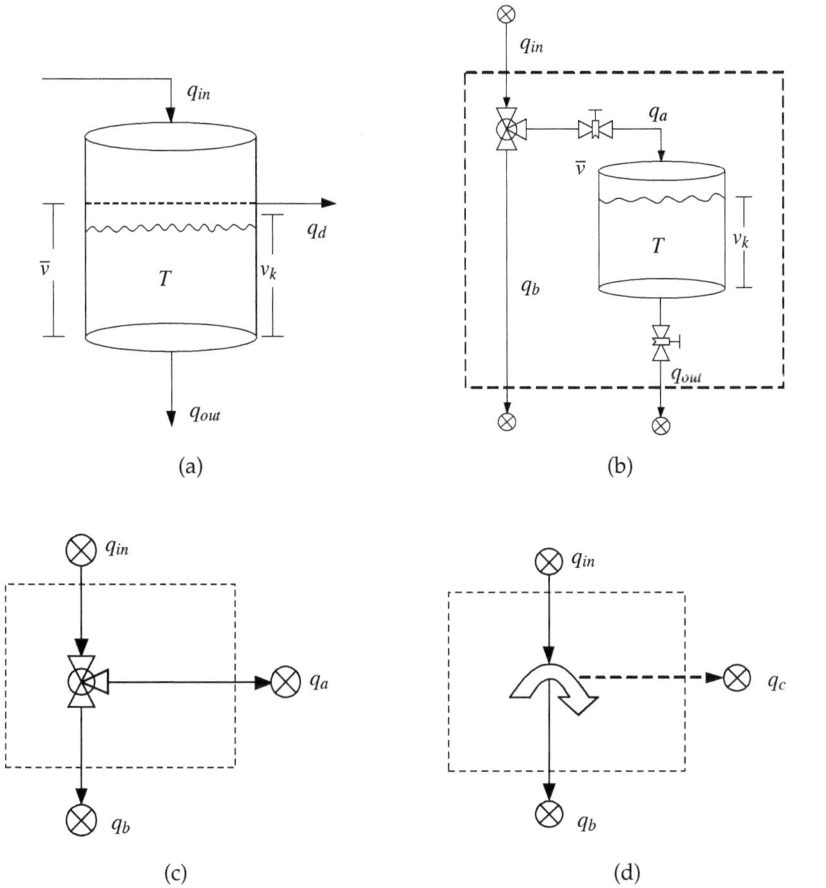

Figure 6.8 Schemes for constitutive elements of sewer networks: (a) virtual tank, (b) real tank, (c) redirection gate, and (d) sewage pipe or weir with a single inflow

Real Tanks

The following expressions related to the tank inflow and outflow are used for this element:

$$q_{out\,k} = \text{sat}\,(q^{\star}_{out\,k}, \beta\, v_k)\,, \tag{6.27}$$

$$q_{ak} = \text{sat}\left(q^{\star}_{ak}, \min\left(\frac{\overline{v}_k - v_k}{\Delta t}, q_{in\,k}\right)\right). \tag{6.28}$$

Thus, the difference equation for the tank volume is again written as in (6.26):

$$v_{k+1} = v_k + \Delta t (q_{ak} - q_{out\,k}), \tag{6.29}$$

and the flow q_b obeys the mass balance

$$q_{bk} = q_{in\,k} - q_{ak}. \tag{6.30}$$

Redirection Gates

For redirection gates, the PWLF model is defined by noting that q_a should satisfy the restriction (5.32), which can be rewritten in terms of the PWL functions as

$$q_{ak} = \mathrm{sat}(q_{ak}, q_{i}n_k). \tag{6.31}$$

The flow through q_b is given by the mass balance in (6.30).

Sewage Pipes (or Single-inflow Weirs)

The PWLF model for both of these elements can be obtained from the overflow condition as follows:

$$q_{bk} = \mathrm{sat}(q_{in\,k}, \overline{q}_b), \tag{6.32}$$

$$q_{ck} = \mathrm{dzn}(q_{in\,k}, \overline{q}_b), \tag{6.33}$$

where $\overline{q}_b$ again corresponds to the maximum flow capacity of the nominal outflow sewer main.

6.5.2 Simulations and Results

This section focuses on comparing the performances of MPC controllers whose prediction models were obtained using either the PWLF modelling approach or the hybrid systems modelling approach presented in Chapter 5. This comparison is performed by considering the set of real-rain episodes in Table 3.3. The computation times required by these two modelling approaches are also compared. The results of this comparison illustrate the benefits of using the modelling approach proposed in this section for RTC implementation in the case study of a real network. The assumptions made for all the implementations will be presented and their validity discussed before the results are given.

Remark 6.4. In order to get an adequate idea of the complexity of the hybrid problem, the system model in MLD form can be equivalently represented in PWA form as stated in Heemels *et al.* (2001). Hence, the system can be expressed as

$$x(k+1) = A_i\, x(k) + B_i\, u(k) + f_i, \quad i = 1, \ldots, N_d, \qquad \text{for} \quad \begin{bmatrix} x(k) \\ u(k) \end{bmatrix} \in \Omega_i, \tag{6.34}$$

where $x(k) \in \mathbb{R}^{n_c}$ is the vector of n_c system states (tank volumes in the network), $u(k) \in \mathbb{R}^{mi_c}$ corresponds to the vector of the mi_c control inputs (manipulated flows), f_i are real vectors associated with the model disturbances (rain), and N_d denotes the number of different dynamics.

According to Bemporad (2004), the way that the PWA system is obtained from the MLD form is based on enumerating all 2^{r_ℓ} combinations of binary variables, where r_ℓ is the number of Boolean variables in the MLD model. For the case study of this book, the MLD model has $r_\ell = 22$, which leads to $N_d = 4194304$ different dynamical modes. Since the complexity associated with the size of the PWA model was high enough, the alternative – to use this representation for the MPC design – was not considered. ◊

MPC Controller Set-up and Implementation

Simulations of the BTC in a closed loop with an MPC law based on a PWLF model of the system were performed. The same parameters and controller set-up as those in the simulations presented in Chapter 5 for hybrid system models were considered.

The PWLF model of the system is associated with a nonlinear optimisation problem. By replacing the nonlinear equality constraints (deriving from the definition of the PWLF model) in the objective function (5.64), a new optimisation problem with a PWLF-based objective function and bounding constraints is obtained. The algorithm used to solve this problem was chosen after evaluating several of the solvers available in Tomlab® (*e.g.*, *conSolve*, *nlpSolve*, among others). The *Structured Trust Region* (STR) algorithm (Conn *et al.*, 1995) was finally chosen since it provides an *acceptable* trade-off between system performance and computational burden. If the global optimum is desired, the previously mentioned nonlinear algorithm should be combined with a bisection approach, as suggested in Boyd and Vandenberghe (2004).

Control Performance and Computation Time Comparisons

For performance comparison purposes, and in addition to the control results obtained when the different modelling approaches (hybrid and PWLF models) are used, performance results from when the network is placed in an open loop are also presented. The computation times reported in this section were obtained using MATLAB® 7.2 implementations running on an Intel® Core™2, 2.4 GHz machine with 4Gb RAM. Note that the computation time results reported here for hybrid models are different from those presented

Table 6.3 Performance results for flooding (all performance values in hm^3)

Rain episodes	Open loop	PWLF model	Hybrid model
1999-09-14	108	88.2	92.9
2002-10-09	116.1	113.3	97.1
2002-07-31	160.3	132.8	139.7
1999-10-17	0	0	0
2000-09-28	1	1	1

Table 6.4 Performance results for pollution (all performance values in hm^3)

Rain episodes	Open loop	PWLF model	Hybrid model
1999-09-14	225.8	226.1 (1.16%)	223.5
2002-10-09	409.8	407.7 (2.25%)	398.7
2002-07-31	378	380 (1.44%)	374.6
1999-10-17	65.1	59.9 (3.09%)	58.1
2000-09-28	104.5	102 (4.08%)	98

Table 6.5 Performance results for sewage treated at WWTPs (all performance values in hm^3)

Rain episodes	Open loop	PWLF model	Hybrid model
1999-09-14	278.4	276.7 (1.43%)	280.7
2002-10-09	533.8	534.2 (1.98%)	544.9
2002-07-31	324.4	321.9 (1.80%)	327.8
1999-10-17	288.4	293.5 (0.61%)	295.3
2000-09-28	285.3	287.5 (1.51%)	291.9

in Table 6.1, due to differences in the characteristics of the machines and the solver versions used. However, the results are still representative in terms of performance indices.

As stated in previous simulations, the open-loop case consists of the sewage system without control, so the manipulated links are used as passive elements. In other words, the flows q_{u1}, q_{u2} and q_{u4} only depend on the inflow to the corresponding gate; they are not manipulated, while q_{u3} is the outflow from the real tank due to gravity (tank discharge).

Control performance results are summarised in Tables 6.3, 6.4 and 6.5 for five of the representative rain episodes presented in Table 3.3. Table 6.3 shows a comparison of the volumes of sewage released to the street (flooding), while Table 6.4 shows the same comparison but for volumes sent to receiving environments (pollution). Finally, Table 6.5 shows a comparison of the volumes of sewage treated at the WWTPs.

Tables 6.3, 6.4 and 6.5 indicate that the hybrid modelling approach generally provides better system performance than the PWLF-based modelling approach. Also note that the inprovement in performance is mostly related to the improvement in the main control objective; there is less improvement

Table 6.6 Computation time results (all time values in s)

Rain episodes	PWLF model		Hybrid model	
	Total CPU time	Max. CPU time for a sample	Total CPU time	Max. CPU time for a sample
1999-09-14	695.33	91.32	1109.29	787.17
2002-10-09	293.23	66.01	561.73	85.31
2002-07-31	830.20	83.04	1050.54	381.49
1999-10-17	180.22	16.15	79.14	10.39
2000-09-28	120.88	12.13	84.76	13.27

in the second objective, and so on. Thus, the pollution produced by the PWLF-based modelling approach during some episodes is worse than it is in the open-loop case; see, *e.g.*, episodes 1999-09-14 and 2002-07-31 in Table 6.4. However, the PWLF-based modelling approach gives the best performance indices for flooding during both episodes, which is in line with the pre-established prioritisation control objectives .

In general, these results were expected, since the MPC controller based on the hybrid modelling approach achieves its optimum by solving a set of convex linear problems through a branch-and-bound scheme. However, the MPC based on the PWLF model leads to a nonlinear network model representation, which may result in a quasiconvex optimisation problem. Therefore, when the STR algorithm is used, the global optimum cannot be assured, as a bisection method is implemented. This could lead to a sequence of suboptimal control actions when the MPC law is computed, which explains why the performance obtained using the PWLF model was generally worse than that obtained when using the hybrid model. The suboptimality levels of the results obtained using the PWLF model were never greater than 4.1% for the second and third objectives (as shown in Tables 6.4 and 6.5 in brackets). The results for the first control objective (related to flooding) were not as homogeneous as one of the modelling approaches gives better system performance in some scenarios while the opposite may be true for other scenarios.

On the other hand, the main difference between using the hybrid or the PWLF modelling approaches is the computation time required to determine the control actions at each iteration. As mentioned in Chapter 5, the MLD form of the model contains a significant number of Boolean and auxiliary variables. Remember that in the BTC case study, determining the control actions using an HMPC controller implies that, for each time instant, and with $H_p = 6$, a set of $2^{22\times 6} = 5.4 \times 10^{39}$ LP problems (for a linear norm in the cost function) or QP problems (for a quadratic norm in the cost function) should be solved in the worst case. Thus, the complexity of the MIP problem associated with the MPC law is enhanced as the number of Boolean variables increases, as the underlying optimisation problem is combinatorial and $\mathcal{NP}$-hard (Papadimitriou, 1994). Thus, the worst-case computation time is exponential in the number of integer variables.

In sewer networks, the number of elements with switching dynamics can change with the topology of the particular case study. Therefore, computation times increase until a point is reached where it is almost impossible to use this modelling approach to obtain an MPC-based RTC law. On the other hand, the alternative modelling approach based on the PWLFs proposed in this section allows control sequences to be computed in less time, but at a price: some degree of suboptimality due to a possible local optimum. Table 6.6 summarises the computation times achieved by both of the modelling approaches proposed in this chapter when they were applied to the rain episodes considered. It can be seen for the last two rain episodes that the maximum computation times for both modelling approaches are almost equivalent. This is due to the influence of system disturbances on the optimisation problem, according to the discussion presented in Section 6.1.

Depending on the variability of the rain and its intensity as a function of the location, more switching behaviours may be observed. This can determine the relative complexity of the MIP problem, but it does not influence the optimisation problem based on a PWLF model in the same way. When a hybrid model is used, many dynamical modes can change due to the evolution of the rain intensity, making the global behaviour of the system more complex (*e.g.*, new flows can appear, the Hamming distance between the Boolean variable vectors at a particular time instant and the previous step increase, *etc.*). This means that the MIOP also becomes more complex, and implies a high computational burden. On the other hand, when a PWLF model is applied to the same rain intensity conditions, the computation does not depend on Boolean variables (which determine the current mode of the system in the hybrid model) and so the computational burden is reduced, even though the optimisation is nonlinear.

To conclude, despite the suboptimal nature of the solutions – which is a consequence of the improved control performance – the MPC controller in the PWLF-based modelling approach not only computes control sequences more quickly but also provides feasible control sequences from the perspective of the real-time restriction imposed by the sampling time. On average, all of the maximum times taken to compute the MPC control action when the PWLF-based modelling approach is used are less than a third of the sampling time. This is not the case when using the hybrid modelling approach.

6.6 Summary

Due to the high CPU times obtained when the BTC was simulated in a closed loop using an HMPC controller (Chapter 5), it is easy to conclude that the time spent computing the HMPC control law is very high in some cases.

The computational cost has been shown to be strongly influenced by the values of the initial conditions at each time instant $k \in \mathbb{Z}_+$. Furthermore, the results obtained indicate that the computation time required to determine the HMPC law is very difficult to predict when an associated MIP problem needs to be solved.

Given this computation time issue, it is necessary to explore some ways to relax and/or simplify the discrete optimisation problem and to find methodologies that make HMPC designs practical when applied to large problems such as the RTC of sewer networks. Most of the hybrid control approaches presented in the literature have been applied to rather small examples. There is no standard strategy for relaxing the problem in order to find a trade-off between optimality and an acceptable amount of computation time for large-scale systems. Some strategies to deal with the MIP problem can be considered before attempting to obtain its solution. Such techniques may simplify the initial hybrid model of the system, split the MIP problem into small subproblems, or add more constraints to the discrete optimisation problem in order to reduce the number of feasible modes, among other approaches.

This chapter proposed a suboptimal model predictive control scheme for a discrete-time hybrid system where optimality is sacrificed in order to reduce computation time. This approach is based on limiting the commutation of the system between its dynamical modes, and takes advantage of the optimal solution computed in the previous sample of the receding horizon strategy. The stability of the proposed scheme, when states are fully measured or accurately estimated and when no disturbances are present, was proven. The proof was based on some of the results reported in Lazar *et al.* (2006), but also incorporated the limitations of switching between dynamical modes.

Once the proposed approach had been explained and discussed under the proviso of no disturbances, it was included in the framework of the HPMC design for sewer networks. Some practical issues were outlined and ways of solving them discussed. It was shown that the use of the suboptimal scheme consistently reduces the computation time as a function of the parameter M, and, in the case of the BTC, the level of suboptimality was not found to be critical.

Additionally, an alternative approach to the design of MPC controllers for large-scale sewage systems was discussed. This utilised a modelling approach based on piecewise linear functions (PWLF). This approach avoids the need to handle Boolean variables (discrete optimisation), but it does include nonlinear functions, resulting in non-convex optimisation problems. However, it does not avoid the problem of finding a feasible solution, and nonlinear programming algorithms only reach local optima, which implies a suboptimal solution (Bazaraa *et al.*, 2006). Considering the well-discussed hybrid systems modelling approach, the control performances and associated computation times obtained when both approaches were used were compared, utilising the BTC as a case study. With the proposed PWLF-based modelling formulation, despite the loss of performance caused by the resulting non-convex optimisation problem, its capacity to reduce the computation time enables this approach to be applied to the RTC of large-scale sewer networks.

Part III
Fault-tolerance Capabilities of Model Predictive Control

Chapter 7
Model Predictive Control and Fault Tolerance

Having presented and discussed the application of the MPC technique to sewer networks in previous chapters, now the occurrence of a fault in an actuator of the sewage system is considered. Therefore, tolerance strategies within the MPC theory and different aspects of the application of a fault-tolerant scheme when MPC is applied as the control law of the closed-loop scheme are discussed, before solving some problems using the proposed methodology. Note that the hybrid modelling methodology developed in Chapter 5 is used to deal with system faults and to model them so that a global solution can be found for the sewer network control problem despite the presence of faults. The expression of faults in the hybrid system framework complements the plant model and allows to take advantage of the capabilities of MPC within an FTC architecture.

7.1 General Aspects

As discussed in Chapter 2, FTC is concerned with the control of faulty systems. In general, control algorithms are designed to achieve control objectives of systems that do not encounter faults. Hence, the presence of a fault would imply changing the control law or even the whole control loop configuration. This method of achieving fault tolerance requires an on-line fault diagnosis scheme and reacts to the the results of the diagnostic algorithms. Another way to achieve fault tolerance is to make use of the robustness of feedback control systems, which leads to implicit fault tolerance. In this case, the control law has been designed to achieve the control objectives whether there are faults in the system or not.

A closed-loop control system can be considered to be fault tolerant if it includes either control law adaptation strategies or redundancy mechanisms for its sensors and/or actuators. Two main groups of control strategies can be distinguished. In a *passive* technique, the control law is designed to take the appearance of a fault into account as a system perturbation. In this case, the controller is set up to achieve the control objectives whether or not faults are present (assuming that this is possible, and with a certain degree of degradation in system performance) using robust control techniques. Thus, to a certain degree, the control law has inherent fault-tolerant capabilities that allow the system to cope with the presence of faults. Works such as Chen *et al.* (1998), Liang *et al.* (2000), Qu *et al.* (2001), Liao *et al.* (2002), Qu *et al.* (2003), Niemann and Stoustrup (2005), Benosman and Lum (2008) and Steffen *et al.* (2009), among many others, contain complete descriptions of passive FTC techniques. On the other hand, in an *active* FTC technique (such as those used in this chapter), the control loop is adapted based on the information provided by a fault detection and isolation (FDI) module within the fault-tolerant architecture (see Blanke *et al.*, 2006). Using this information, some automatic adjustments are made in order to achieve the control objectives if possible. See Zhang and Jiang (2003), Zhang and Jiang (2008) and Benosman (2010) for extensive reviews of active FTC.

A fault can be seen as a discrete event that affects the system by changing some of its particular properties (its structure and/or its parameters). An active FTC scheme should therefore detect and isolate the fault and, if possible, estimate its magnitude (fault diagnosis) via the FDI module, before adapting the controller to the fault situation such that the control objectives are satisfied even in the presence of the fault (control redesign). In particular, the whole active FTC scheme can be expressed via the three-layer architecture for FTC systems proposed by Blanke *et al.* (2006), where the first layer corresponds to the control loop, the second layer corresponds to the fault diagnosis and accommodation modules, while the third layer is related to the supervisory system.

Some important reasons for splitting FTC systems into layers include the clearly defined structure that results, the ability to specify each layer individually, and because it enables the detection and supervisory functions and/or modules to be tested. However, there is no guarantee that the whole FTC system will operate properly when its subsystems/layers are integrated. Due to the discrete-event nature of fault occurrences and the implementation of controller redesign, an FTC system can be considered a hybrid system, which makes its analysis and design nontrivial. This was noted in Parisini and Sacone (1998), where the three levels identified by Blanke *et al.* (2006) were denoted the *continuous-state level* (corresponding to the control level), the *interface level* (associated with the fault diagnosis and accommodation level), and the *discrete-event level* (relating to the supervision level). However, so far, the hybrid nature of FTC systems has been widely ignored in order to facilitate a simple design, reliable implementation, and systematic testing.

One of the main objectives of this chapter is to embed the active fault-tolerant design of controllers based on MPC within the hybrid system framework while considering their dynamical nature. Several approaches to handling hybrid systems are given in the literature (see, *e.g.*, Heemels *et al.*, 2001). In the particular approach proposed here, a hybrid model of a system that includes a given set of fault modes is proposed. An active fault-tolerant HMPC (AFTHMPC) controller is then designed using hybrid system modelling and control methodologies. This allows the HMPC to account for the fault information provided by an FDI module, which automatically results in on-line controller redesign. The information about the fault is the discrete event (discrete input) that switches the mode of the controller when the fault occurs, enabling the controller redesign. In other words, the proposed methodology of FTC design allows to design the three layers of an FTC system in an integrated manner and to verify the global behaviour of the whole FTC scheme using hybrid systems theory.

7.2 Fault-tolerant Control and Hybrid Systems

As discussed before, active FTC requires the inclusion of an on-line FDI module and a supervisory module within the fault-tolerant architecture. When a fault is detected and isolated by the FDI module, the supervisor reacts by activating some remedial actions that allow the faulty control loop to continue to satisfy the given specifications. As discussed in Section 7.1, an architecture for active fault tolerance consisting of three layers was proposed in Blanke *et al.* (2006):

- Layer 1 (*control loop*): this comprises a traditional control loop with sensor and actuator interfaces, signal conditioning and filtering, and the controller. The solution of a control problem involves finding a particular control law among a given set of *control laws* $\mathcal{U}$ that enables the controlled system to achieve the *control objectives* $\mathcal{O}$ while ensuring that its behaviour also satisfies a set of *constraints* $\mathcal{C}$. Thus, the solution of the problem is completely defined by the triple $\langle \mathcal{O}, \mathcal{C}, \mathcal{U} \rangle$.
- Layer 2 (*fault diagnosis and accommodation*): this comprises an FDI module and a closed-loop redesign module, which, in turn, defines the remedial actions performed by the supervisor. The functions of this module are fault detection based on either physical or analytical redundancy using fault detection and isolation methods, the detection of faults in control algorithms, and the application of software and effector modules to execute fault accommodation actions.
- Layer 3 (*supervision*). The supervisor is a discrete-event dynamical system (DEDS) that considers state-event logic in order to monitor the state (mode) of the controlled system, and transitions between the states of the supervisor are induced by events (faults). When the system is in a fault mode, a set of pre-established remedial actions should be activated in order to correct and recover the faulty system and then to achieve the set control objectives.

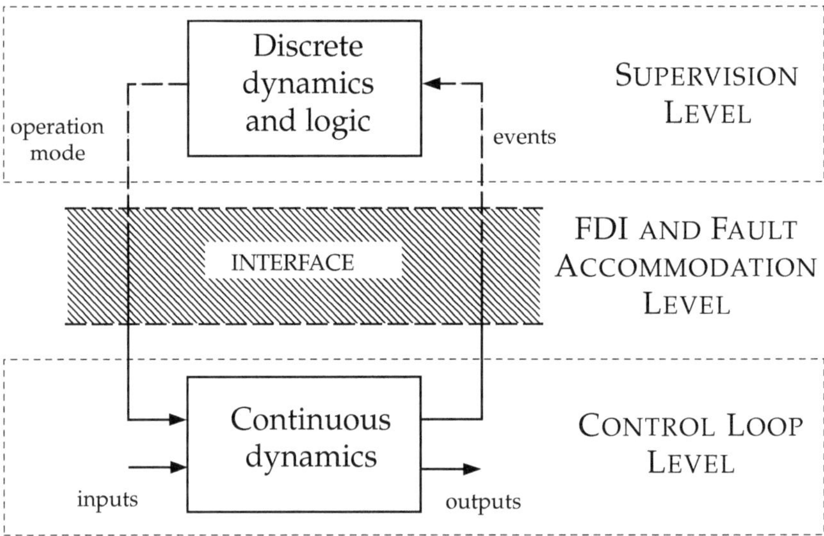

Figure 7.1 Parallelism between the basic scheme of a hybrid system and the three levels of FTC architecture

As noted in Parisini and Sacone (1998), if this structure is compared with the conceptual scheme of a hybrid system (see Figure 7.1), it is apparent that the two schemes are very similar. In particular, the control-loop level corresponds to the continuous part of the hybrid system, while the FDI and redesign level corresponds to the interface between continuous and discrete system dynamics. Finally, the supervision level corresponds to the discrete dynamics part of the hybrid system. Moreover, events correspond to faults within the FTC architecture and redesign actions to changes in the operational mode of the continuous part. This parallelism leads to the use of modelling and control methodologies that can consider both continuous and discrete dynamics. The continuous part is associated with each particular operational mode and can typically be modelled using physical principles. On the other hand, the discrete part derives from logic conditions that establish commutations of operational modes, depending on internal system variables or external events (*e.g.*, faults).

Moreover, faults in control systems can affect any element of the the control loop (sensors, actuators or the plant), thus changing either the structure or parameters. According to Blanke *et al.* (2006), dynamical models of a plant subject to faults must be considered when designing a fault-tolerant controller. The occurrence of faults change the set of constraints of the control problem $\mathscr{C}(\theta)$ by changing either the parameters or the constraints themselves. These changes can be seen as new operational modes associated with the type of the fault that has occurred. In this way, when the system to be

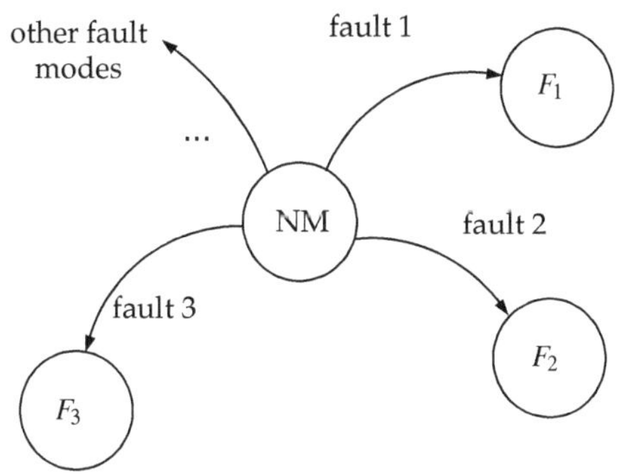

Figure 7.2 Conceptual scheme of changes in the mode of the system due to the effects of faults. *NM* denotes the nominal (fault-free) system mode

controlled is modelled, two types of operational modes should be taken into account (see Figure 7.2):

- *Intrinsic operational modes*, which are related to changes in the system due to the evolution of its hybrid dynamics
- *Faulty operational modes*, which are related to changes in the system due to the occurrence of a fault.

Once the hybrid model of the system that considers the set of operational modes (intrinsic and faulty) has been defined, a control methodology that takes advantage of this information needs to be devised. Note that the total number of intrinsic modes considered depends on the accuracy of the hybrid description of the system; *i.e.*, it is possible to take all hybrid dynamics into account, but this will increase the complexity of the system representation. On the other hand, there will be as many fault modes as defined fault models (considering that only one fault can occur at each time instant). Hybrid MPC strategies which consider models that include operational modes (intrinsic and faulty) are discussed in the following sections.

7.3 Fault-tolerance Capabilities of MPC

7.3.1 Implicit Capabilities

As mentioned before, the robustness of feedback control systems gives rise to a certain degree of implicit fault tolerance. Faults in a closed-loop control scheme are often compensated for by the control action. The same applies when MPC is used at control level. It has also been demonstrated that, even when the knowledge of the fault is not available, if the estimation of external disturbances affecting the closed loop is performed in a convenient way and the input variables are constrained in a hard manner, the MPC controller will automatically take advantage of actuator redundancy when available (Maciejowski, 1998). However, when the states are assumed to be measurable or estimable, this fault tolerance property does not apply.

Information on the occurrence of a fault can be included in an MPC law in the following ways:

- By changing the constraints to represent certain faults; it is especially *easy* to adapt algorithms to faults in actuators
- By modifying the system prediction model of the MPC controller in order to reflect the influence of the fault on the plant
- By relaxing the initial control objectives in order to reflect the system's limitations when operating under fault conditions.

However, these approaches rely on several assumptions (Maciejowski, 1999):

- That the nature of the fault can be located and its effects can be modelled
- That the internal model of the plant can be updated, essentially in an automatic manner
- That the set of control objectives defined during the MPC design process can be left unaltered once the fault has occurred.

These important assumptions can accounted for by using a reliable FDI which takes advantage of emerging technologies that can be employed for system management as well as to make it easier for the designer/user to interact with the complex system.

The utilisation of FTC when MPC is used as the control law has been reported in the literature over the last few years. The first steps in this field were discussed in Maciejowski (1997), and the theory was implemented for an aircraft system in Maciejowski and Jones (2003). The main results reported allow to conclude (among other things) that MPC has a high level of tolerance to some faults – especially actuator faults – under certain conditions, even when the faults are not detected. Since then, MPC and its fault tolerance have inspired lots of contributions in the FTC field; see Patwardhan *et al.* (2006), Abdel-Geliel *et al.* (2006) and Mendonca *et al.* (2006), among others. In Prakash *et al.* (2005), the authors propose a scheme where fault tolerance and MPC work together in chemical applications, as MPC controllers are typically used to control key operations in chemical plants, so this can have important impacts on both the productivity and the safety of the entire system. In the same way, Zumoffen *et al.* (2008) introduce improvements to fault tolerance methods used in the application of an MPC strategy to a complex chemical process. Based on MPC, Zhou (2004) propose an FTC design that considers faults on actuator elements. They present and discuss simple fault detection and fault complement approaches. Deshpande *et al.* (2009) address the design of state estimators in the framework of fault-tolerant nonlinear MPC, while Miksch *et al.* (2008) propose the real-time implementation of fault-tolerant architectures based on predictive control strategies.

7.3.2 Explicit Capabilities

Linear constrained MPC involves solving an optimisation problem using either linear or quadratic programming, which determines the optimal control action. As the coefficients of the linear term in the cost function and the right-hand side of the problem constraints depend linearly on the current state, the quadratic programming in particular can be viewed as a multi-parametric quadratic programming (mpQP) problem. In Bemporad *et al.* (2002b), the authors analyse the properties of mpQP, showing that the optimal solution is a PWA function of the vector of parameters. As a consequence, the MPC controller is a PWA control law, which not only ensures feasibility and stability but is also optimal with respect to LQR performance. An algorithm based on a geometric approach to solving mpQP problems in order to obtain explicit receding horizon controllers was proposed in Bemporad *et al.* (2002b).

The explicit form of the MPC controller also provides additional insight that can yield a better understanding of the control policy. Moreover, this methodology allows faults to be introduced as additional parameters in parametric programming algorithms though the information provided by an FDI module. For instance, in the case of faults that affect actuator bounds, since the maximum control input from an actuator is often constrained in the optimisation problem formulation, this constraint can be considered a parameter. Thus, if an actuator has failed, the situation can be handled by constraining the corresponding control input to be null, or – based on the fault information available – by constraining the related control input to have the new (faulty) operating ranges. For more on explicit solutions for MPC controllers, see Alessio and Bemporad (2009).

7.4 Including Fault Tolerance in HMPC

As discussed in Section 7.1, an FTC system is similar to a hybrid system. Given the fault-tolerant capabilities of the MPC described in this chapter, it is worth using an MPC controller within the three-level architecture discussed previously. This section deals with the interaction of blocks within an FTC scheme that takes into account the hybrid modelling of the plant for MPC purposes.

Looking at the schemes shown in Figures 7.3 and 7.4, which will be used in later sections to explain fault-tolerant HMPC strategies, it is clear that the plant is treated as a hybrid system, where the continuous and discrete parts interact via the event generator interface. External effects such as disturbances and/or noise may affect some parts of the plant in a selective way. Even though faults can occur in any constitutive element within the system, they can be seen as external events that affect its nominal behaviour, altering the system dynamics. Furthermore, the closed-loop controller may send either continuous or discrete inputs to the plant, depending on its nature and the control design. The following sections present a description of fault-tolerant HMPC strategies according to the information available on the influence of faults in a plant described by a hybrid model and the design of the HMPC controller.

7.4.1 Implicit Fault-tolerant HMPC

According to Maciejowski (2002), if fault information is available, either the internal model or system constraints can be modified accordingly. In this way, fault tolerance can be *implicitly* incorporated into an MPC controller in a natural way. Furthermore, due to the flexibility with which the control objectives can be expressed within the MPC formalism, when faults cause control objectives to become unattainable, they can be dropped from the optimisation problem or lowered in priority, for example by changing hard constraints to soft ones (Kerrigan *et al.*, 2000).

However, looking at the FTC in the hybrid framework, it is possible to consider a set of fault modes when the hybrid model for prediction is determined. This model considers not only the intrinsic dynamical modes of the system but also the added fault modes. This addition of fault modes can also cause changes in the control problem constraints.

When fault modes are included in the hybrid model used by the HMPC controller, the implementation of active fault tolerance is very straightforward. When a fault is known to have occurred (due to the existence of an FDI module), a change of mode is induced in the prediction model at the controller. This mode change allows the hybrid controller to adapt its operational mode to handle the faulty plant operational mode. In what follows, this strategy is referred yo as *active fault-tolerant hybrid model predictive control* (AFTHMPC). Figure 7.3 depicts a conceptual scheme of this strategy.

In this chapter, it is assumed that the modules associated with the fault-tolerant mechanisms are not affected by faults since they are designed properly. However, in practice, faults may occur in these modules, thus adding more uncertainty to the whole fault-tolerant architecture. Moreover, the FDI module is assumed to work correctly (although this is an important condition); that is, the fault is immediately detected and isolated after its occurrence.[1] Moreover, if the FDI module reports the magnitude of the fault,

[1] This can be easily guaranteed for abrupt faults through the appropriate design of the FDI module. On the other hand, fault detection and isolation of incipient faults can be delayed, or they may even remain undetected in some cases. An undetected fault should be treated as an uncertainty, and the FTC problem should be addressed using a passive approach.

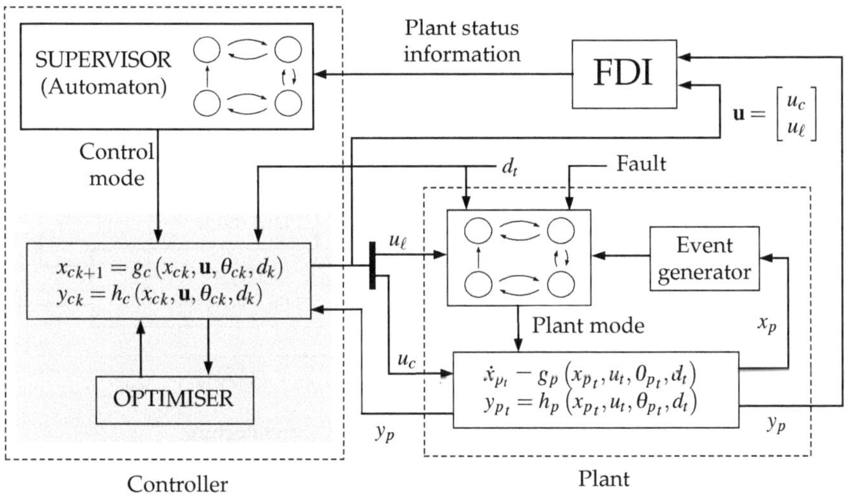

Figure 7.3 Scheme of the AFTHMPC architecture

fault accommodation should then be possible. Otherwise, the whole control scheme should be *reconfigured*.

Definition 7.1 (Fault Accommodation (Blanke *et al.*, 2006)). Accommodating the system to the fault consists of solving the control problem $\langle \mathscr{O}, \hat{\mathscr{C}}_f(\hat{\theta}_f), \hat{\mathscr{U}}_f \rangle$, where $\hat{\mathscr{C}}_f(\hat{\theta}_f)$ is an estimate of the actual set of system constraints and parameters provided by the FDI module. In other words, fault accommodation refers to the adaptation of the set of controller parameters to the dynamical properties of the faulty plant; *i.e.*, the system under the influence of the fault.

Definition 7.2 (Control Reconfiguration (Blanke *et al.*, 2006)). Reconfiguring the control law once the fault occurs involves finding a new set of constraints $\mathscr{C}_f(\theta_f)$ such that the control problem $\langle \mathscr{O}, \mathscr{C}_f(\theta_f), \mathscr{U}_f \rangle$ can be solved. Once found, the resulting control problem is solved and the reconfigured control law is applied. This approach also considers whether fault accommodation cannot be implemented. It should be noted that the determination of the reconfigured control signals strongly depends on the nature of the faults detected, and the new control law must be redesigned on-line.

When the presence of the fault is not detected, the hybrid controller needs to deal with a plant that has changed its operational mode due to the effect

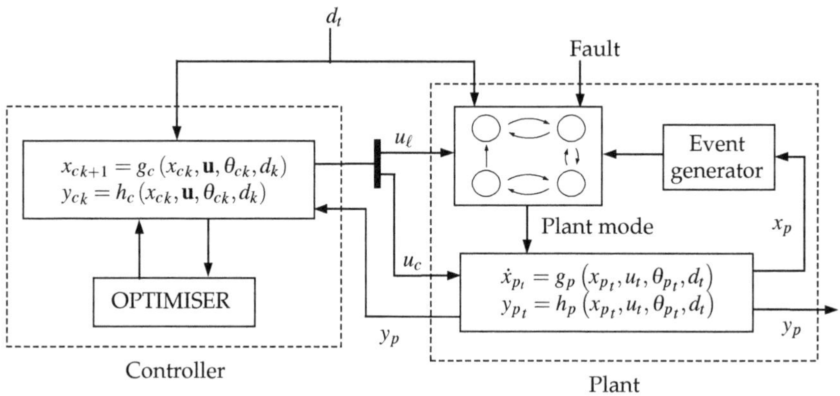

Figure 7.4 Scheme of the PFTHMPC architecture

of the fault. In this case, fault tolerance relies on the inherent capabilities of the feedback control loop. This is referred to as *passive fault-tolerant hybrid model predictive control* (PFTHMPC) (see Figure 7.4). In this case, the conceptual scheme is the same as that for AFTHMPC but without the FDI and supervisor modules. In practice, this means that the HMPC controller should be designed to be robust to the uncertainty introduced by the fault.

When the hybrid plant model including the fault modes is represented using the MLD formalism, the MLD form introduced in (2.8) is modified as follows:

$$x(k+1) = Ax(k) + B_1u(k) + B_2\delta(k) + B_3z(k) + B_f f(k), \tag{7.1a}$$

$$y(k) = Cx(k) + D_1u(k) + D_2\delta(k) + D_3z(k) + D_f f(k), \tag{7.1b}$$

$$E_2\delta(k) + E_3z(k) \leq E_1u(k) + E_4x(k) + E_5 + E_f f(k), \tag{7.1c}$$

where f represents the faults modelled as Boolean inputs, assuming that only information about their existence is provided in real-time by the FDI module. If additional knowledge about the fault is available, such as its magnitude, a more detailed fault model can be included in the MLD representation given by (7.1). Depending on the case, matrices B_f, D_f, and E_f may contain values related to the magnitude of the fault rather than just its presence. Therefore, the HMPC problem should be modified accordingly by replacing the MLD form in (2.14) (2.8) with the MLD form that includes faults in (7.1).

Trying to define all of the fault modes while accounting for the influence of each fault could be extremely difficult, and it is sometimes impossible. In practice, fault tolerance is only provided for a set of given fault modes (Blanke *et al.*, 2006). These are therefore the only fault modes that should be included in the hybrid model of the plant. Moreover, this set of faults constitutes the specification for the FDI module, since this module should be able to detect and isolate each particular fault from this given set.

Notice that a new MLD (or PWA) system is generated for each of the fault modes considered. This implies that the terminal constraint set $\mathbb{X}_T$ and the cost function guaranteeing stability should be changed to those determined, for instance, by the computational method provided in Lazar *et al.* (2006). The task of changing these conditions and the model constraints affected by the fault can be implemented in several ways, as will be discussed in Section 7.5 below.

7.4.2 Explicit Fault-tolerant HMPC

The approach presented in Bemporad *et al.* (2002b) provides an explicit form of a linear MPC controller that performs the computations needed to implement MPC off-line while preserving all of its other characteristics. Such an explicit form of linear MPC is based on the solution of a QP problem in which the coefficients of the linear term in the cost function and the right-hand side of the constraints depend linearly on the current state. The QP problem associated with linear MPC can therefore be seen as an mp-QP problem. The optimal solution is a PWA function of the vector of states (parameters). As a consequence, the solution is an explicit MPC law that is PWA with respect to the states. This result can be generalised to hybrid systems by transforming the MLD system given in (2.8) and in the HMPC optimisation problem (2.14) into an equivalent PWA system (Bemporad, 2004). In particular, Borrelli *et al.* (2005a) described a way to transform the HMPC control law (2.14) into an explicit PWA form by introducing the faults as additional states, which extends the state vector as

$$\tilde{x}(k) = \begin{bmatrix} x(k) \\ f_1(k) \\ f_2(k) \\ \vdots \\ f_q(k) \end{bmatrix}, \tag{7.2}$$

where f_i, $i \in \{1,2,\cdots,q\}$ corresponds to the considered faults (and is also related to the number of fault modes). The controller can be updated based on the fault information provided by the FDI module as follows:

$$u_{\mathrm{MPC}}(\tilde{x}(k)) = \begin{cases} F_1\tilde{x}(k) + g_1 & \text{if} \quad \tilde{x}(k) \in \Omega_1 = \{\tilde{x}(k) : H_1\tilde{x}(k) \leq S_1,\} \\ & \vdots \\ F_{s+q}\tilde{x}(k) + g_{s+q} & \text{if} \quad \tilde{x}(k) \in \Omega_{s+q} = \left\{\tilde{x}(k) : H_{s+q}\tilde{x}(k) \leq S_{s+q}\right\}, \end{cases}$$

where $\cup_{i=1}^{s+q}\Omega_i$ is the set of states (s is the number of intrinsic modes) and faults for which a feasible solution to (2.14) exists.

In this way, the optimisation problem (2.14) tackled with the MLD model including faults (7.1) can be computed off-line rather than on-line.

7.4.3 An Illustrative Example

In order to illustrate the concept of fault-tolerant explicit MPC, it will assumed that faults affect the operating range of the actuator of a given plant. Consider a first-order continuous system described by the transfer function

$$G(s) = \frac{0.8}{2s+1},$$

which, using a sampling time $\Delta t = 0.1$s, has an equivalent discrete-time state-space description that is given by

$$\begin{aligned} x_{k+1} &= 0.9512x_k + 0.0975u_k, \\ y_k &= 0.8x_k, \end{aligned} \tag{7.3}$$

where $|u_k| \leq \mu$ and $\mu = 1$. For notation proposes, $u_k \in [\underline{u}, \overline{u}] = [-1, 1]$. An MPC controller is designed by considering the following cost function:

$$J(x_k, u_k) = Px_{H_p}^2 + \sum_{i=0}^{H_p-1} \left(Qx_i^2 + Ru_i^2\right), \tag{7.4}$$

where $H_p = 2$ (for simplicity of presentation) and the terminal weight matrix P is determined using the discrete-time algebraic Riccati[2] equation (DARE), with $Q = 1$ and $R = 0.1$ (Lancaster and Rodman, 1995). According to Theorem 6.2.1 in Goodwin *et al.* (2005), and given the value for H_p, the explicit form of the optimal control law $u_k^* = \mathscr{K}_2(x)$, which depends on the current system state $x_0 = x$, is given by[3]

$$\mathscr{K}_2(x) = \begin{cases} -\text{sat}_\mu(Gx + h) & \text{if} \quad x \in \mathbb{Z}^-, \\ -\text{sat}_\mu(Kx) & \text{if} \quad x \in \mathbb{Z}, \\ -\text{sat}_\mu(Gx - h) & \text{if} \quad x \in \mathbb{Z}^+, \end{cases} \tag{7.5}$$

where the saturation function $\text{sat}_\mu(\cdot)$ is defined, for the saturation level μ, as

[2] The mathematician Jacopo Francesco Riccati (1676–1754) was born in Venice and is remembered for his famous studies of ordinary differential equations.

[3] Results used to show the fault tolerance capabilities of MPC analytically have limited application considering that $H_p = 2$. However, results related to mp-programming problems are more general.

$$\text{sat}_{\mu}(u_k) = \begin{cases} \mu & \text{if} \quad u_k > \mu, \\ u_k & \text{if} \quad |u_k| \leq \mu, \\ -\mu & \text{if} \quad u_k < -\mu. \end{cases} \tag{7.6}$$

Matrices K and P are obtained through the DARE

$$P = A^T PA + Q - K^T (R + B^T PB)K, \tag{7.7}$$

$$K = (R + B^T PB)^{-1} B^T A, \tag{7.8}$$

which results in $P = 3.2419$ and $K = 2.2989$. Also from Theorem 6.2.1 in Goodwin *et al.* (2005), the gain $G \in \mathbb{R}^{1 \times n}$ and the constant $h \in \mathbb{R}$ are given by

$$G = \frac{K + KBKA}{1 + (KB)^2} \quad \text{and} \quad h = \frac{KB}{1 + (KB)^2}\mu, \tag{7.9}$$

which results in $G = 2.6557$ and $h = 0.2135$. The state-space partitions for control law (7.5) are defined by

$$\begin{aligned} \mathbb{Z}^- &= \{x : K(A - BK)x < -\mu\}, \\ \mathbb{Z} &= \{x : |K(A - BK)x| \leq \mu\}, \\ \mathbb{Z}^+ &= \{x : K(A - BK)x > \mu\}, \end{aligned} \tag{7.10}$$

which, in turn, determines the following sets:

$$\begin{aligned} \mathbb{Z}^- &= \{x : 1.6713x < -\mu\}, \\ \mathbb{Z} &= \{x : 1.6713x \leq \mu\}, \\ \mathbb{Z}^+ &= \{x : 1.6713x > \mu\}. \end{aligned}$$

Bearing in mind that $\overline{u} = \mu$ and $\underline{\mu} = -\mu$, it is apparent that control law (7.5) indirectly depends on the actuator limits through expressions in (7.9) and (7.10). Therefore, it is clear how the effect of a fault in the actuator operating range can modify the expression of the control law. This suggests that (7.5) can be parameterised as a function of the actuator faults (limits). This parameterisation can be performed based on the results reported in Bemporad *et al.* (2002b), where an explicit state-feedback control law for the MPC controller, which is PWA with respect to the states, can be derived through a mpQP problem statement (Tøndel *et al.*, 2003). Hence, using (7.3), consider an extended state-space system written as

$$\begin{bmatrix} x(k+1) \\ \overline{\mu}(k+1) \\ \underline{\mu}(k+1) \end{bmatrix} = \begin{bmatrix} 0.9512 & 0 & 0 \\ 0 & 0 & 1 \\ 0 & 1 & 0 \end{bmatrix} \begin{bmatrix} x(k) \\ \overline{\mu}(k) \\ \underline{\mu}(k) \end{bmatrix} + \begin{bmatrix} 0.0975 \\ 0 \\ 0 \end{bmatrix} u(k), \tag{7.11}$$

$$\begin{bmatrix} y(k) \\ y_{\overline{\mu}}(k) \\ y_{\underline{\mu}}(k) \end{bmatrix} = \begin{bmatrix} 0.8 & 0 & 0 \\ 0 & 0 & 1 \\ 0 & 1 & 0 \end{bmatrix} \begin{bmatrix} x(k) \\ \overline{\mu}(k) \\ \underline{\mu}(k) \end{bmatrix} + \begin{bmatrix} 0 \\ -1 \\ -1 \end{bmatrix} u(k). \tag{7.12}$$

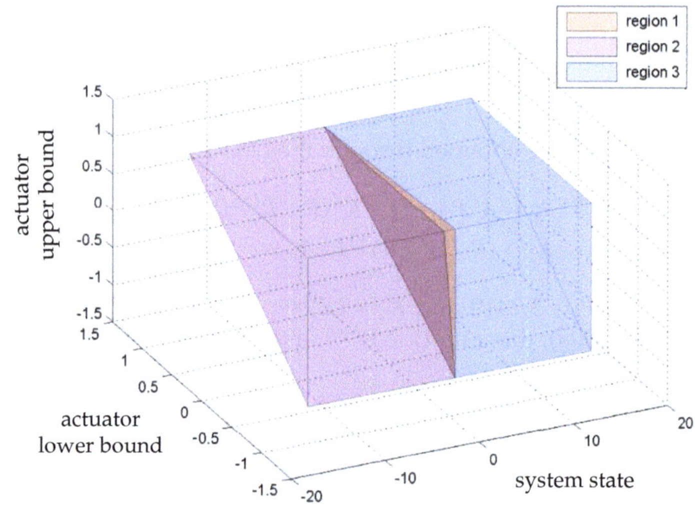

Figure 7.5 Polyhedral partitions of the control law (7.13)

The expression of $\mathscr{K}_2(\cdot)$, given as a function of $\theta = [x\ \overline{\mu}\ \underline{\mu}]^T$, corresponds to the explicit PWA control law given by

$$\mathscr{K}_2(\theta(k)) = \begin{cases} \begin{bmatrix} -2.299 & 0 & 0 \end{bmatrix} \theta(k) & \text{if } \begin{bmatrix} 1.6713 & 1 & 0 \\ -1.6713 & 0 & -1 \end{bmatrix} \tilde{x}(k) \leq \begin{bmatrix} 0 \\ 0 \end{bmatrix}, \\ & \text{(Region \#1)} \\ \begin{bmatrix} -2.656 & 0 & -0.2135 \end{bmatrix} \theta(k) & \text{if } \begin{bmatrix} 0 & 1 & -1 \\ 1.272 & 0 & 0.7611 \end{bmatrix} \theta(k) \leq \begin{bmatrix} 0 \\ 0 \end{bmatrix}, \\ & \text{(Region \#2)} \\ \begin{bmatrix} -2.656 & -0.2135 & 0 \end{bmatrix} \theta(k) & \text{if } \begin{bmatrix} 0 & 1 & -1 \\ -1.272 & -0.7611 & 0 \end{bmatrix} \theta(k) \leq \begin{bmatrix} 0 \\ 0 \end{bmatrix}, \\ & \text{(Region \#3)} \end{cases} \tag{7.13}$$

which was obtained using multi-parametric programming algorithms included in the Hybrid Toolbox for MATLAB® (Bemporad, 2006). Figure 7.5 shows how the optimal control gain varies with the system state and the faults considered (the upper/lower limits of the actuation range). To summarise, if a fault in the actuator range appears, the use of an explicit MPC controller (7.13) allows on-line adaptation based on the fault estimate provided by the FDI module.

7.5 Some FTHMPC Implementation Schemes

In this section, once implicit and explicit ways of including fault tolerance in HMPC have been presented, various methods of implementing AFTHMPC using the MLD formulation will be summarised and discussed. First of all, there are two ways in which the controller can handle the fault modes of the plant:

- Internally, by including the fault modes in the MLD model of the system in the same way as for intrinsic modes. This approach generates additional Boolean variables corresponding to the fault modes, which should be initialised according to the fault information provided by the FDI module. In this case, the automata-based supervisor is embedded in the global MLD model of the system.
- Externally, by generating a different MLD model for every possible fault mode. An external automata-based supervisor system then selects the MLD model that should be used at each iteration.

Additionally, the HMPC controller for each MLD model can be computed:

- On-line (implicitly), by solving the optimisation problem associated with the HMPC controller at each iteration, or
- Off-line (explicitly), by transforming the MLD model into a PWA model and then computing a PWA controller.

Given that there are two ways of handling fault modes in the MLD model and two ways of computing the HMPC controller for a given MLD model, there are four implementation options, as shown in Figures 7.6 and 7.7. The option depicted in Figure 7.6 (a) is the most straightforward, since it simply includes the fault modes in the hybrid model of the system by using the MLD formalism, so that they coexist with the inherent hybrid modes. The controller will commute between both modes on-line, depending on external (faults) or internal (changes of operational mode) events. The stability of the control loop can be guaranteed by setting some design conditions appropriately, just as in the traditional MPC design process, such as the length of the prediction horizon, terminal state constraints, *etc.* (Maciejowski, 2002). In this architecture, the classical supervisor module is embedded in the HMPC model. In Figure 7.6 (b), only hybrid modes are included in the MLD model used by the HMPC controller. This means that a bank of MLD models (one for each fault mode) needs to be designed, and an external supervisor should change the MLD model used by the controller depending on the external fault event.

Each MLD model is obtained by considering the fault modes within the nominal hybrid dynamics of the plant, thus resulting in the same number of MLD models as fault modes considered. If both architectures are compared, the first provides a complete hybrid view of the FTC problem, while the second one follows the classical three-level architecture. Both architectures

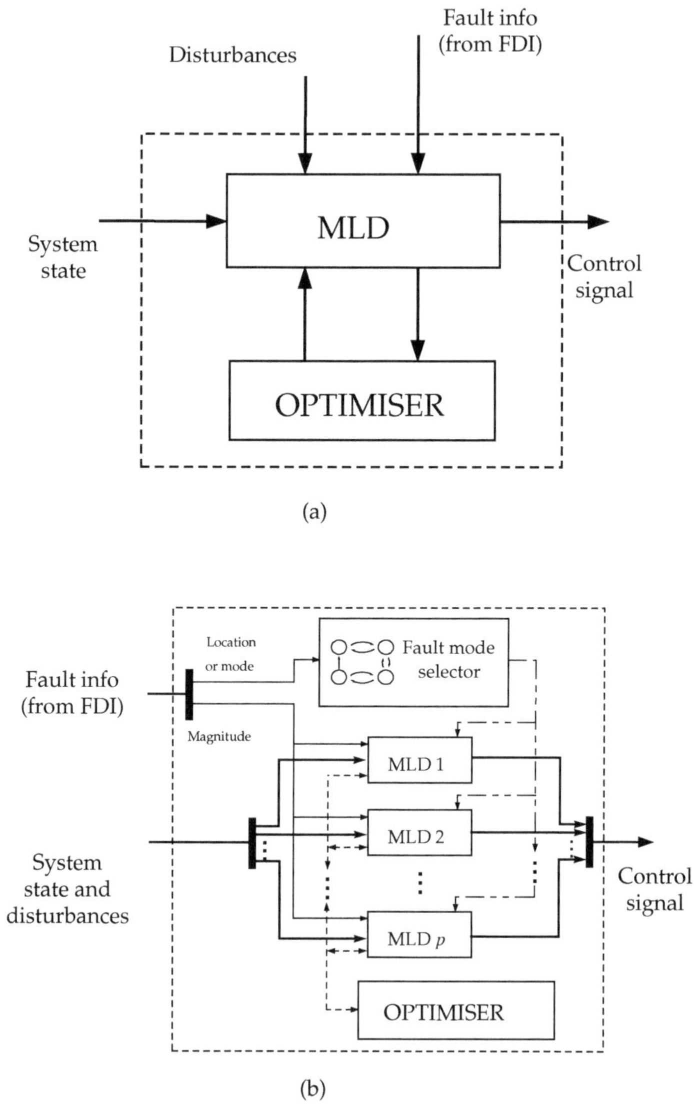

Figure 7.6 FTHMPC implementation architectures with optimisation-based controllers

are based on the implicit fault-tolerant capabilities of the MPC strategy, as described in previous chapters. The benefit of the first option is that external design of the supervisor is not needed, but the price of this is that additional Boolean variables must be included to reflect the fault modes.

On the other hand, if the explicit computation of the HMPC controller is considered, two additional architectures can be designed. In Figure 7.7

(a), an explicit MPC law that is parameterised with respect to system states and faults is proposed. The advantage of this architecture is that it utilises an off-line controller that includes pre-established reconfiguration logic. The reconfiguration mechanism associated with a particular fault is derived in the same way as the mechanism for a change of mode. The drawback of this approach is that the number of regions of the resulting PWA controller grows very quickly with the number of states and faults considered. One possible way to reduce this complexity is to generate an explicit MPC controller for each fault mode (Figure 7.7 (b)). This implies the use of an external supervisor that switches between controllers, and which incorporates the conditions needed to ensure closed-loop stability (Maciejowski, 2002; Lazar *et al.*, 2006).

7.6 Fault-tolerant HMPC of Sewer Networks

In this book, the FTHMPC strategy for sewer networks considers that actuator faults change the bounds of the operating ranges of the manipulated flows $\underline{q}_{u_i}$ and $\overline{q}_{u_i}$. Information about such changes is made available once the FDI has detected, isolated and estimated the actuator fault in the sewage system.

FTHMPC is now implemented and tested on the BTC, in line with the ideas discussed in previous sections. This section considers typical fault scenarios related to control gates of sewer networks and discusses the modelling of these phenomena within the hybrid system framework. The use of the hybrid approach is motivated by the results obtained in a previous study where faults were considered within a linear MPC design. In this section, some obtained results considering different FTHMPC strategies are presented. As mentioned before, the FDI module in the AFTHMPC strategy is assumed to operate properly. It detects and isolates the faulty actuator and returns the corresponding information along with the magnitude of the fault. The fault information is assumed to be readily available, and is used to modify the corresponding constraints in the discrete optimisation problem.

7.6.1 Fault Scenarios

There are many types of control gate fault scenarios that can occur within a sewer network. In this chapter, three fault scenarios are considered, since they represent typical phenomena that occur in faulty control gates. The low end of the flow range can be limited by an inability to close a gate, or the high end of the range can be limited by an inability to open a gate sufficiently (or a reduction in pump capacity if pumping elements are considered). A

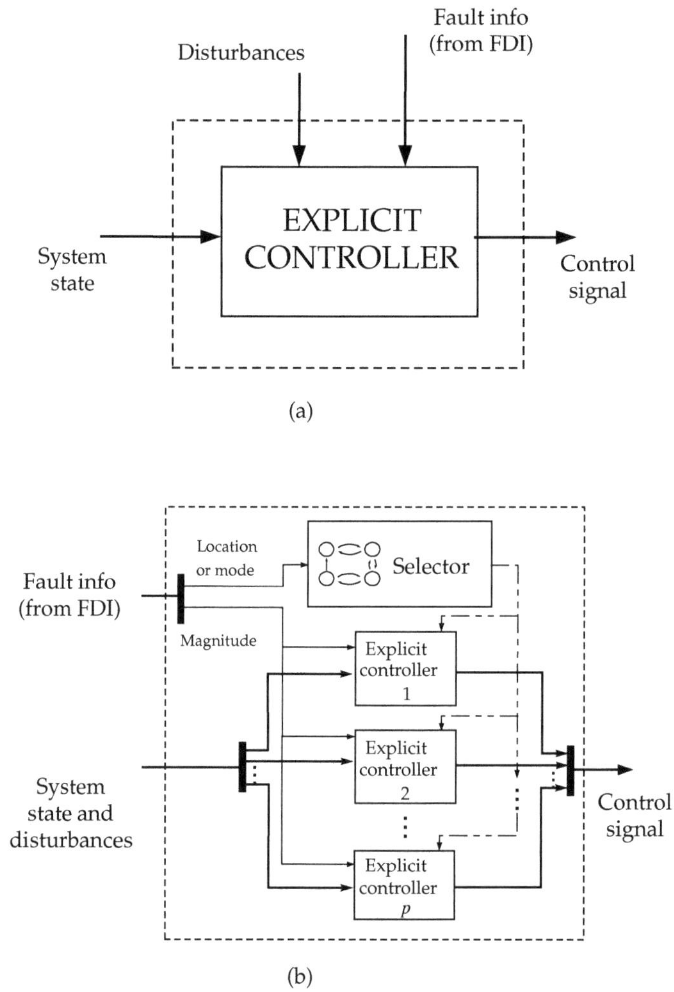

Figure 7.7 FTHMPC implementation architectures with explicit-solution-based controllers

stuck gate means that the range is limited to a point or very narrow interval. Hence, control gate fault scenarios involve limiting the flow ranges of the gates in the following three ways:

1. Limit the low end of the range (so that the range is 50–100%), which is denoted $f\underline{q}_{u_i}$
2. Limit the high end of the range (so that the range is 0–50%), which is denoted $f\overline{q}_{u_i}$
3. Limit both ends of the range, thus simulating a stuck gate (so that the range is 50–51%), which is denoted $f\underline{\overline{q}}_{u_i}$.

In scenarios 1 and 3, $\underline{q}_{u_i} = 0.5\overline{q}_{u_i}$. For the BTC in Figure 5.6, and particularly in scenario $f\underline{q}_{u_2}$, the lower limit of the manipulated outflow from gate C_2, namely $\underline{q}_{u_2}$, is set equal to the upper limit of the corresponding outflow from the actuator C_3, $\overline{q}_{u_3}$. The reason for this arrangement is that the optimisation problem is unfeasible if tank T_2 is full and $\underline{q}_{u_2} > \overline{q}_{u_3}$.

7.6.2 Linear Plant Models and Actuator Faults

When the prediction model is linear, the AFTMPC strategy must deal with a optimisation problem that has a convexity which depends on the type of fault considered. Assuming that no fault modifies the upper limit of any actuator range, the optimisation problem can be solved using fast LP or QP algorithms, depending on the expression of the cost function. This approach can therefore deal with big systems with many state variables, so relatively big optimisation problems can be solved quickly, resulting in an optimal global solution. Note that, under these conditions, the problem is readily scalable in the sense of the size of the sewer network.

However, when a fault causes the range of at least one system actuator to adopt a non-zero lower bound, the constraint related to this actuator becomes non-convex, and so the optimisation problem also becomes non-convex. In this case, it is not possible to take advantage of the LP or QP algorithms to solve the optimisation problem, and its solution is not global (*i.e.*, it is suboptimal). This problem was reported in Ocampo-Martinez *et al.* (2005), where a small system inspired by the BTC was used. The proposed system contains representative elements of the whole sewer network of Barcelona and considers enough components to be able to test the FTC strategies based on MPC. The results obtained shown the utility of the fault-tolerant approach when certain models of faults are considered, including models where $\underline{q}_{u_i}$ is non-zero. Nevertheless, for fault models that modify $\underline{q}_{u_i}$, the solution obtained corresponded to a local minimum in the cost function, which implies that the system performance is suboptimal.

7.6.3 Hybrid Modelling and Actuator Faults

The ideas proposed in Ocampo-Martinez *et al.* (2005) and the discussion in Section 7.4 compels to explore other ways of modelling the fault scenarios considered. Hence, this chapter uses the hybrid system modelling presented in Chapter 5 to account for actuator faults in sewer networks by considering fault scenarios to be modes that are related to the behaviour of the element. Note that the system changes between modes depending on the fault sce-

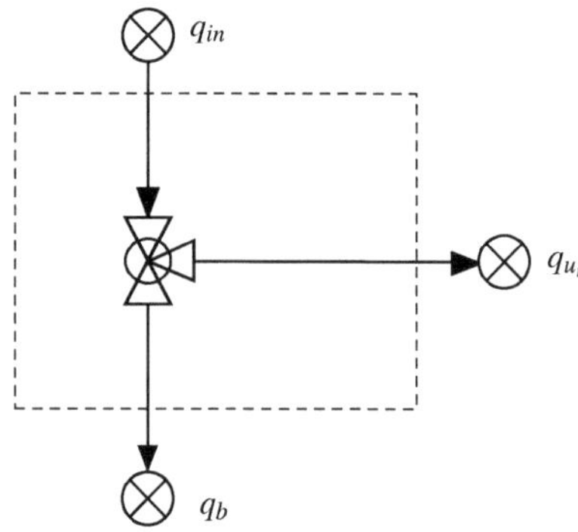

Figure 7.8 Control gate scheme used to explain the hybrid modelling of faults

nario. In this case, it will assumed that a single fault affects the system. Two or more faults occurring at the same time will cause an explosion in the number of system modes.

For instance, consider the fault actuator mode $f\underline{q}_{u_i}$ for the redirection gate presented in Figure 5.3, which is repeated here for simplicity. In this case, it will be shown how such a mode can be expressed using the proposed hybrid modelling approach.[4]

This fault explicitly limits the range of the manipulated outflow q_{u_i}. The following expressions are therefore obtained using the principle of mass conservation:

$$q_{u_i} \leq \min(\overline{q}_{u_i}, q_{in}), \tag{7.14a}$$
$$q_{u_i} \geq \min(\underline{q}_{u_i}, q_{in}). \tag{7.14b}$$

The presence of these expressions implies the addition of non-convex constraints to the optimisation problem. Note that this behaviour is similar to the switching behaviour associated with weirs, and it suggests the appearance of new system modes. This fact again encourages to use the proposed hybrid modelling methodology.

Inequality (7.14a), related to the upper bound, can be expressed using two linear inequalities as

$$q_{u_i} \leq \overline{q}_{u_i}, \qquad q_{u_i} \leq q_{in}.$$

Note that, in the case of fault scenarios $f\overline{q}_{u_i}$ and $f\underline{q}_{u_i}$, the fault affects the system when $q_{in} > f\overline{q}_{u_i}$. Otherwise, the fault does not have any influence on the behaviour of the network element.

[4] The other fault modes described in Section 7.6.1 can be modelled in a similar manner using this proposed approach.

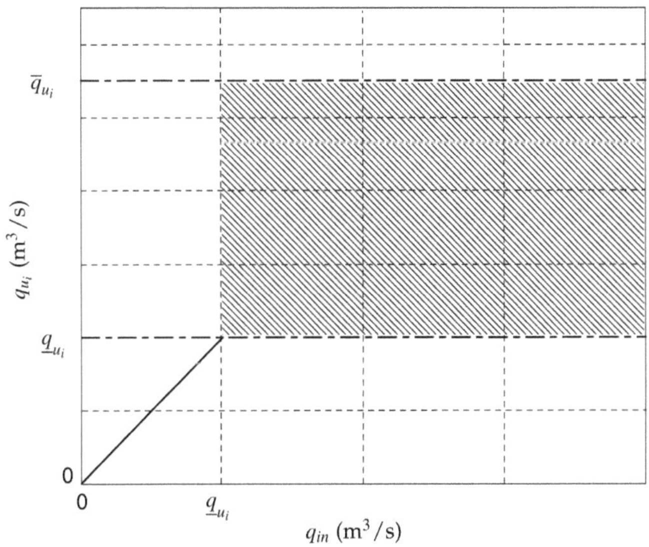

Figure 7.9 q_{u_i} values for which the actuator constraints are fulfilled

The non-convex constraint (7.14b) can easily be treated in the hybrid framework by introducing the auxiliary variables

$$[\delta_i = 1] \rightarrow [q_{u_i} \leq q_{in}], \tag{7.15a}$$

$$z_{q_{u_i}} = \begin{cases} \underline{q}_{u_i} & \text{if} \quad \delta_i = 1, \\ q_{in} & \text{otherwise}, \end{cases} \tag{7.15b}$$

and rewriting (7.14b) as $q_{u_i} \geq z_{q_{u_i}}$. In this way, a non-convex constraint can be expressed via a finite number of linear constraints (using the equivalences (5.1) and (5.2)), thus avoiding possible problems resulting from the convergence of optimisation routines to local minima. Figure 7.9 shows the set of q_{u_i} values for which the actuator constraints are fulfilled. Note the change in the area upon the modification of the bound $\underline{q}_{u_i}$ under the influence of the fault. Also note that points over the line defined by $q_{u_i} = q_{in}$ and (7.14a) belong to this set of valid values when $\underline{q}_{u_i}$ is non-zero.

This fault modelling associated with the hybrid approach for sewer network elements suggests that each element can include typical fault models that depend on the role of the element within the network. Hence, the fault tolerance is included in the modelling process when the plant model for the MPC controller design is obtained, assuming that the fault information is provided by an FDI module. This feature leads to a more accurate representation of the real plant when the effect of a given fault is considered, and makes it easier to either accommodate the fault or reconfigure the control law within the FTC architecture. Another advantage of the hybrid modelling of system faults is the possibility of having continuous values for the magnitude of the fault, given its particular dynamical model. This also increases the accuracy of the element model when a fault occurs.

7.6.4 Implementation and Results

The purpose of this section is to compare the results of applying AFTHMPC and PFTHMPC to the BTC in Figure 5.6 during realistic rainstorm episodes (Section 3.3.3) and actuator faults (Section 7.6.1). The assumptions made in this comparison will be presented and their validity discussed before the results are given. In all cases, fault accommodation was performed. For this particular case study, the reconfiguration strategy would involve considering the control gate to be completely open when a actuator fault occurs. Thus, the sewage flows downstream due to gravity, fulfilling the mass conservation principle and respecting the main paths. However, note that this behaviour occurs when fault scenarios with a non-zero lower bound on the operational range are considered. This observation is valid while the gate inflow is lower than the faulty lower bound of its outflow.

Simulating the Scenarios

The control strategies were compared by simulating the closed-loop system for all of the fault scenarios presented in Section 7.6.1 and for the rain episodes listed in Table 3.3. The structure of the closed loop was the same as that used for the simulations in Chapter 5. In the PFTHMPC case, the actuator ranges were limited only in the plant model (simulation model).

The durations of the simulation scenarios were selected to be approximately 8 h ($k \in [0, 100]$), as the rainstorms generally had peaks lasting around ten samples or 50 min. The tanks were empty at the beginning of each scenario. In order to compare strategies, the total flooding, the pollution and the treated sewage released into the environment were each summed across the whole scenario.

Horizons H_p and H_u were selected to be six samples or 30 min for the reasons given in previous chapters. The cost function structure, norms and control tuning were the same as those used in the simulations performed for nominal HMPC, as shown in Chapter 5.

Main Results

In general, street flooding was reduced by a larger amount when AFTMPC was used rather than PFTMPC. The greatest improvements were obtained when the precipitation was large enough to force the actuators to operate close to the upper limits of their ranges (*i.e.*, when the precipitation pushed the sewer network close to its capacity). Even though results are shown for specific rain episodes, the conclusions presented here were based on simulations of various scenarios.

AFTMPC did not yield great improvements when heavy rain episodes such as the one that occurred on September 14, 1999 (see Figure 3.7 (a)) were considered in the simulation. The reduction in CSO achieved was about 0–5%. The reason for this is that the BTC does not have the capacity to handle rainstorms of such an intensity, even in the fault-free case. Therefore, it did not matter whether the actuation limits were known to the HMPC controller or not. This behaviour is presented numerically in Table 7.1, where the main performance indices obtained for the BTC when using AFTHMPC and PFTHMPC are compared for the fault scenarios considered. Note that the largest reduction in flooding was obtained for scenarios $f\overline{q}_{u_2}$ and $f\overline{\underline{q}}_{u_2}$. In those scenarios, the flooding was reduced from roughly 135,700 m^3 to around 121,000 m^3, which corresponds to an improvement of about 11%. The other performance indices were also improved. However, sometimes the flooding was reduced but the pollution and/or treated sewage were not. This is due to the prioritisation of the control objectives, which is reflected in the tuning of the cost function.

When very common rain episodes with little precipitation were studied, the same results were observed; *i.e.*, applying AFTMPC did not result in any great improvement. This is because the constraints are usually not reached in such scenarios, so actuator faults rarely affect system performance.

The results shown in Table 7.2 relate to a rainstorm that occurred on October 17, 1999. This rain episode has a 0.7-year return period with regard to the total amount of rain and a ten-year return period with regard to the maximum rain intensity. The main feature of this episode is its behaviour during the time window considered. As seen in Figure 3.7 (b), this rain event exhibits a double peak in intensity, so the resulting sewer network behaviour is quite complex and the nominal HMPC and the FTHMPC designs have to expend a lot of effort in attempting to control the system and avoiding the effects of a fault. The network is almost unsusceptible to faults at gates C_1 and C_4. The modified upper bounds for outflows q_{u_1} and q_{u_4} were always greater than the inflows of the corresponding control gates. Due to restrictions on the lower bounds of the flow at these gates caused by the faults, the values of the manipulated flows q_{u_1} and q_{u_4} took the same values as their respective inflows in most cases. Those values were not big enough to cause overflows downstream.

Table 7.1 Results obtained when FTHMPC was applied to the BTC during the rain episode that occurred on September 14, 1999 (all performance values are in hm^3)

Fault scenario		PFTHMPC			AFTHMPC		
Actuator	Type	Flooding	Pollution	TS	Flooding	Pollution	TS
q_{u_1}	$f\underline{q}_{u_1}$	99.5	223.5	280.6	99.5	223.5	280.6
	$f\overline{q}_{u_1}$	93.6	223.9	280.3	93.4	223.8	280.3
	$f\overline{\underline{q}}_{u_1}$	99.9	223.9	280.3	99.9	223.9	281.6
q_{u_2}	$f\underline{q}_{u_2}$	92.9	222.7	281.5	92.9	222.7	281.5
	$f\overline{q}_{u_2}$	135.7	230.2	274.1	121.0	228.8	275.5
	$f\overline{\underline{q}}_{u_2}$	125.5	226.6	277.6	118.3	227.5	276.7
q_{u_3}	$f\underline{q}_{u_3}$	94.3	226.0	278.2	94.2	225.3	278.9
	$f\overline{q}_{u_3}$	97.7	221.1	283.2	95.1	223.1	281.0
	$f\overline{\underline{q}}_{u_3}$	97.6	223.1	281.2	96.0	224.7	279.5
q_{u_4}	$f\underline{q}_{u_4}$	102.1	222.4	281.8	102.1	222.3	281.9
	$f\overline{q}_{u_4}$	92.8	223.5	280.7	92.8	223.5	280.7
	$f\overline{\underline{q}}_{u_4}$	102.1	222.4	281.8	102.1	222.3	281.9

Table 7.2 Results obtained when FTHMPC was applied to the BTC during the rain episode that occurred on October 17, 1999 (all performance values are in hm^3)

Fault scenario		PFTHMPC			AFTHMPC		
Actuator	Type	Flooding	Pollution	TS	Flooding	Pollution	TS
q_{u_2}	$f\underline{q}_{u_2}$	0.0	61.9	291.4	0.0	61.9	291.4
	$f\overline{q}_{u_2}$	15.9	62.8	290.5	10.4	63.3	290.0
	$f\overline{\underline{q}}_{u_2}$	14.8	63.3	290.0	9.6	63.8	289.5
q_{u_3}	$f\underline{q}_{u_3}$	0.0	59.1	294.2	0.0	58.8	294.5
	$f\overline{q}_{u_3}$	3.5	57.7	295.6	0.2	58.7	294.7
	$f\overline{\underline{q}}_{u_3}$	0.8	58.8	294.6	0.0	58.9	294.4

In this case, the most notable reduction in flooding occurred in the fault scenario $f\overline{q}_{u_2}$, where an improvement of approximately 35% resulted from the employment of the AFTHMPC strategy compared to PFTHMPC. Figure 7.10 shows how the flow in q_{14}, the overflow in q_{24} and the volume T_3 in the BTC evolved following the second rain peak when active and passive approaches were applied. In the active case, because the manipulated flow q_{u_2} has lost capacity, the controller cannot take advantage of the the real tank T_3

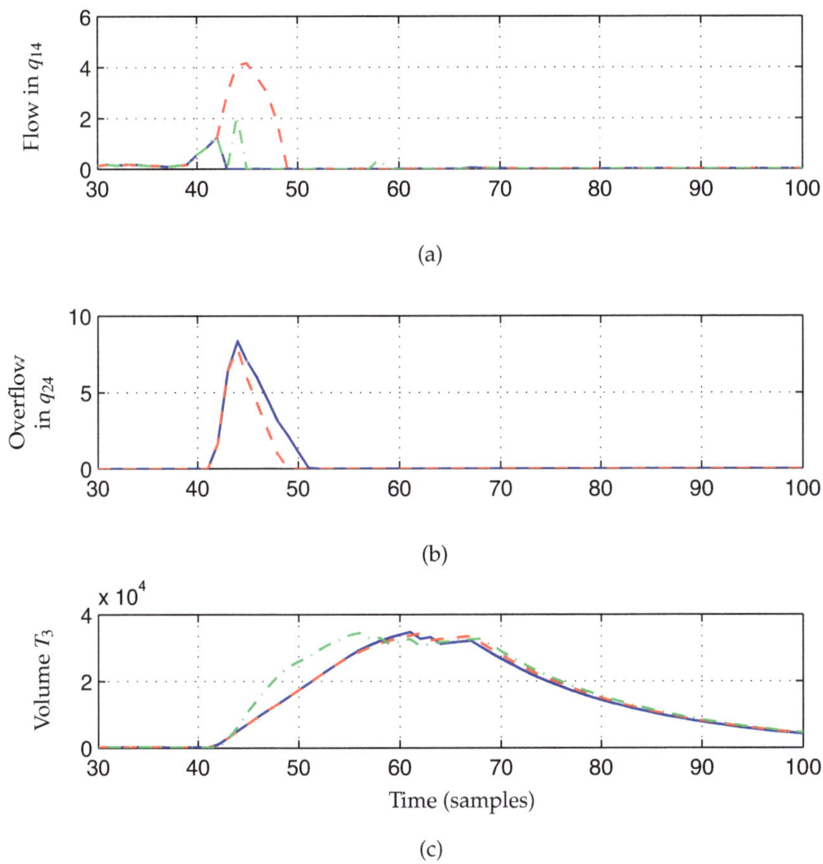

Figure 7.10 Behaviours of different parts of the BTC when active and passive FTHMPC approaches were applied to fault scenario $f\overline{q}_{u_2}$ for rain episode 99-10-17: (a) flow in q_{14}, (b) overflow in q_{24}, and (c) volume T_3. *Solid curve*, PFTHMPC; *dashed curve*, AFTHMPC; *dot-dash curve*, no-fault HMPC

in the short/medium term.[5] This means that sewage from T_1 is conveniently directed through sewer q_{14} (see Figure 7.10 (a)), which in turn means that the sewer mains close to T_3 do not have as much overflow as they do in the passive strategy (see Figure 7.10 (b)). The slow rate of fill of T_3 in conjunction with its convenient outflow manipulation (control signal related to q_{u_3}, see Figure 7.10 (c)) give it a useful buffering capacity, which aids attempts to avoid an overflow in T_5. All of these actions combine to reduce the flooding in this fault scenario for this rain episode.

[5] Bear in mind that real tanks (reservoirs) are generally used as buffers within a network. When used in this way, they can store enough sewage to avoid flooding and/or CSO downstream.

Finally, consider an intermediate type of rain episode in the sense of rain intensity, for instance the one that occurred on September 3, 1999. This episode is well supported by the design of the network topology; *i.e.*, by implementing an adequate control law, it can be ensured that the sewer network does not suffer from flooding. Simulation results obtained for this rain episode indicate that the network is almost unsusceptible to all of the considered fault scenarios at the control gates C_1 and C_4, for the same reasons as those discussed for the previous rain episode. Table 7.3 collects together the results obtained for the other two actuators; these show that AFTHMPC yielded improvements over the PFTHMPC strategy. In the scenarios $f\overline{q}_{u_3}$ and $f\underline{q}_{u_3}$, the FTC strategy achieved a reduction in flooding of around 100%. This improvement was realised because AFTHMPC takes advantage of the accumulation of sewage in the real tank caused by the restricted ability of this tank to empty – an effect of the fault (see Figure 7.11 (b)), and adequately computes the set of control signals needed to redirect the sewage, thus avoiding the presence of large quantities of sewage in the faulty elements of the sewer network.

When the PFTHMPC strategy is used, the controller computes a control signal $q_{u_3}(k)$ without knowing the fault in the actuator, which implies that the computed control signal and the applied signal relating to the control action will differ (note that the applied signal corresponds to the computed signal but is saturated due to the faulty upper limit). Hence, since the physical constraints impose a limit on inflow to the real tank (the manipulated link q_{u_2}) as a function of its current volume, the sewage inflow at C_2 is directed through the sewer main q_{24}, causing an overflow in this element and an increase in flooding.

On the other hand, the AFTHMPC strategy computes the control signal $q_{u_1}(k)$ in such a way that the sewage from T_1 is redirected through q_{14}, so less sewage heads toward the real tank and its faulty output actuator. Thus, despite the slow emptying of T_3, the C_2 sewage inflow is conveniently distributed between sewer mains q_{u_2} and q_{24}, avoiding any overflow in this latter pipe and thereby preventing an increase in flooding.

Figure 7.11 shows the evolutions of the computed flow $q_{u_3}(k)$, the volume in T_3 and the overflow in q_{24} when active and passive FTC strategies were employed.

7.7 Summary

This chapter introduced concepts and methods relating to the incorporation of fault tolerance into a closed-loop scheme governed by an MPC control law. The implicit and explicit fault-tolerance capabilities of this control technique were outlined, and particular features of them were discussed. Moreover, MPC designs that utilize hybrid system models were included

Table 7.3 Results obtained when FTHMPC was applied to the BTC during the rain episode on September 3, 1999 (all performance values are in hm^3)

Fault scenario		PFTHMPC			AFTHMPC		
Actuator	Type	Flooding	Pollution	TS	Flooding	Pollution	TS
	$f\underline{q}_{u_2}$	0.0	44.3	232.3	0.0	44.3	232.3
q_{u_2}	$f\overline{q}_{u_2}$	15.2	44.5	232.1	12.2	44.7	231.9
	$f\underline{\overline{q}}_{u_2}$	14.7	44.3	232.3	11.8	44.4	232.2
	$f\underline{q}_{u_3}$	0.0	45.2	231.4	0.0	45.2	231.4
q_{u_3}	$f\overline{q}_{u_3}$	4.1	44.1	232.5	0.2	44.3	232.3
	$f\underline{\overline{q}}_{u_3}$	1.5	44.3	232.3	0.0	44.3	232.3

in the framework of FTC. A proposed parallelism between the conceptual structure of a hybrid system and the three-level FTC architecture was highlighted. This proposal states that both of these conceptual schemes correspond in terms of signal types and the exchange of status information between modules.

For the hybrid MPC theory, two FTC strategies for dealing with faults were proposed. They differed in the information available to the controller regarding the effects of faults throughout the plant. The strategies proposed follow the philosophy discussed in Chapter 2 for control strategies in an FTC architecture.

Moreover, this chapter presented a comparison between the results of applying AFTHMPC and PFTHMPC to sewer networks during realistic rain and fault scenarios. The results showed that AFTHMPC reduced flooding in almost all cases. Furthermore, AFTHMPC was able to prevent flooding or to reduce it considerably provided the intensity of the rain episode was within the design limits of the sewer network. However, the performance was barely improved for heavy rainstorms due to topological limitations of the sewer network. In other cases, such as light rains, the fault scenarios considered here do not exert much of an influence on the behaviour of the sewer network due to the small internal flows handled.

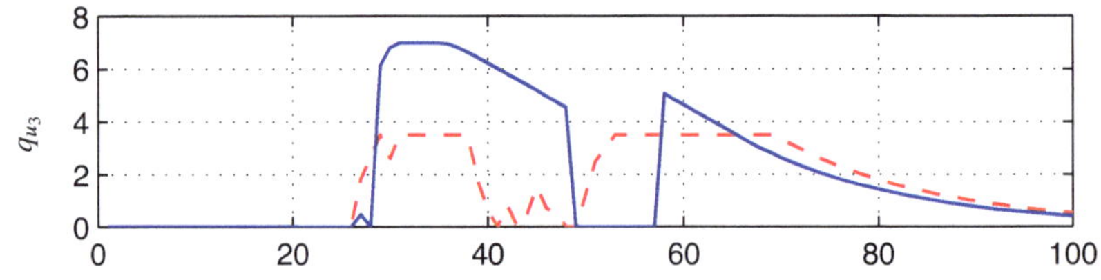

(a)

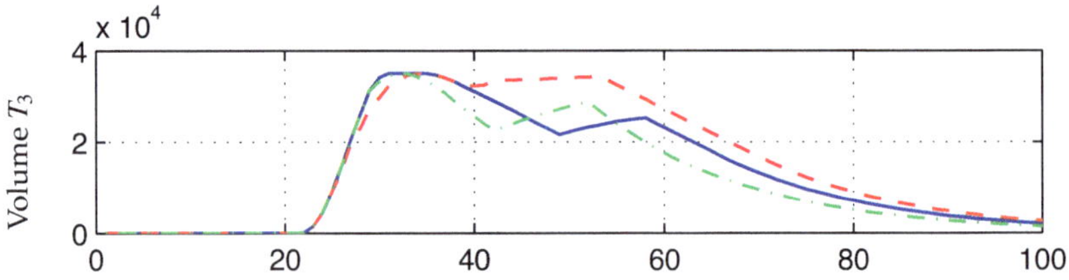

(b)

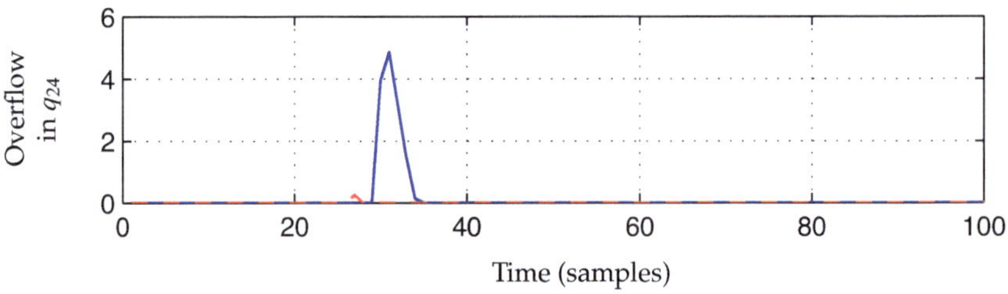

(c)

Figure 7.11 Stored volumes in the real tank T_3 for fault scenario $f\overline{q}_{u_3}$ during rain episode 03-09-1999: (a) $q_{u_3}(k)$, (b) volume in T_3, and (c) overflow in q_{24}. *Solid curve*, PFTHMPC; *dashed curve*, AFTHMPC; *dot-dash curve*, no-fault HMPC

Chapter 8
Fault-tolerance Evaluation of Actuator Fault Configurations

8.1 Introduction

In this chapter, the fault tolerance of a certain actuator fault configuration (AFC) is evaluated within the framework of a linear MPC law with constraints. The issue of fault-tolerance evaluation has already been treated in the literature for an LQR problem without constraints (Staroswiecki, 2003), due to the existence of an analytical solution. However, constraints (on states and control signals) are always present in real industrial control problems, and are easily handled using linear constrained MPC (LCMPC) (Maciejowski, 2002; Rawlings and Mayne, 2009). On the other hand, an analytical solution for obtaining these control laws does not exist, which makes it difficult to apply this approach.

The approach proposed in this chapter are of a computational rather than analytical nature. It follows the idea proposed by Lydoire and Poignet (2004) in which the computation of the control law for a predictive/optimal controller with constraints can be divided into two steps: the computation of a set of solutions that satisfies the constraints (feasible solutions), and then the determination of the optimal solution.

Faults in actuators can cause important changes in the set of feasible solutions, since faults can modify the constraints on the control signals. This means that the set of admissible solutions for the given control objectives may be empty. Therefore, the admissibility of the control law when an actuator fault occurs can be determined if the feasible solutions set is known. One of the aims of this chapter is to provide methods to compute this set and then evaluate the admissibility of the control law.

A CSP could be formulated to find the feasible solutions set for the LCMPC problem (Lydoire and Poignet, 2004). However, such problems are computationally demanding, which means that it is necessary to look for a approximate solution bounded by *interval hulls*[1] in an iterative manner with respect to time. Proceeding in this way, an interval simulation problem is solved implicitly. However, such problems are associated with difficulties such as the *wrapping effect* (Puig *et al.*, 2003). This effect appears when the set of possible states is approximated by a single interval box, causing any overestimation to accumulate with simulation time, thus resulting in an "explosion" of uncertainty. In order to avoid those problems, the region of possible states could be approximated using more complex domains than intervals, such as subpavings (Kieffer *et al.*, 2002; DiLoreto *et al.*, 2007), ellipsoids (ElGhaoui and Calafiore, 2001; Polyak *et al.*, 2004; Sanyal *et al.*, 2008), or zonotopes (Khn, 1998; Alamo *et al.*, 2005; Ingimundarson *et al.*, 2008), among others.

Therefore, this chapter presents a preliminary study of the previously mentioned evaluation of AFC admissibility within the framework of linear plant models. This study can be extended to cases where nonlinear or linear hybrid models are considered. First approximations for nonlinear systems are reported in Ocampo-Martinez *et al.* (2006a), while those for the hybrid system framework are reported in Torrisi (2003).

8.2 Preliminary Definitions

Consider the MPC problem defined in Section 2.2.1, and in particular the sequences $\mathbf{u}_k$ in (2.2), $\mathbf{x}_k$ in (2.4), and

$$\tilde{\mathbf{x}}_k = \left(\mathbf{x}_k \cup \{x_{0|k}\}\right) \setminus \{x_{H_p|k}\}. \tag{8.1}$$

These lead to the following definitions.

Definition 8.1 (Feasible Solutions Set). The feasible solutions set consists of the pairs $(\tilde{\mathbf{x}}_k, \mathbf{u}_k)$ that satisfy the system constraints; *i.e.*,

$$\Omega = \{\tilde{\mathbf{x}}_k, \mathbf{u}_k \;:\; \mathbf{x}_{k+1} = g(\tilde{\mathbf{x}}_k, \mathbf{u}_k)\},$$

where g denotes the system mapping of the states and control inputs.

Without any loss of generality, consider an objective function J defined as

$$J(\mathbf{x}_k, \mathbf{u}_k) = \phi(x_{H_p|k}) + \sum_{i=0}^{H_p-1} \Phi(x_i, u_i), \tag{8.2}$$

[1] Roughly speaking, the interval hull of a closed set $\mathbb{M}$ can be defined as the smallest box that contains $\mathbb{M}$.

where ϕ is a function that constrains the final state value over H_p and Φ is a function of the system states and inputs.

Definition 8.2 (Set of Feasible Control Objectives). The set of feasible control objectives is given by

$$\mathbb{J}_\Omega = \{J(\mathbf{x}_k, \mathbf{u}_k) \in \mathbb{R} \, : \, (\tilde{\mathbf{x}}_k, \mathbf{u}_k) \in \Omega\},$$

and corresponds to the set of all values of a cost function J evaluated in a given feasible solutions set.

In Definition 8.2, when a fault scenario is considered, the set Ω becomes Ω_f and the set $\mathbb{J}_\Omega$ becomes $\mathbb{J}_{\Omega_f}$.

Definition 8.3 (Admissible Solutions Set). Given the following subsets:

- Ω_f, defined as the feasible solutions set related to a particular AFC, and
- $\mathbb{J}_\mathbb{A}$, defined as the set of admissible control objectives,

the admissible solutions set is given by

$$\mathbb{A} = \left\{ \mathbf{x}_k, \mathbf{u}_k \in \Omega_f \, : \, J(\mathbf{x}_k, \mathbf{u}_k) \in \mathbb{J}_\mathbb{A} \cap \mathbb{J}_{\Omega_f} \right\},$$

and corresponds to the feasible solutions subset that produces control objectives in $\mathbb{J}_\mathbb{A}$. If $\mathbb{A}$ is an empty set, then the AFC considered is not admissible.

Definition 8.4 (Predicted States Set). Given the set of states at time *k-1*, the set of predicted states at time k is defined as:

$$\mathbb{X}_k^p = \{x_k = g(x_{k-1}, u_{k-1}) \, : \, x_{k-1} \in \mathbb{X}_{k-1} \wedge u_{k-1} \in \mathbb{U}\},$$

and corresponds to the set of states at time k produced by the evolution of the system starting from the set of states at time *k-1*.

Definition 8.5 (Set of Compatible States (in Prediction)). The set of compatible states at time k is given by

$$\mathbb{X}_k^c = \{x_k \, : \, x_k \in \mathbb{X}_k^p \cap \mathbb{X}\},$$

and corresponds to the set of predicted states compatible with the system state constraints.

Definition 8.6 (Compatible Inputs Set). The set of compatible inputs at time *k-1* is given by

$$\mathbb{U}_{k-1}^c = \left\{u_{k-1} \in \mathbb{U} \, : \, (x_k = g(x_{k-1}, u_{k-1})) \in \mathbb{X}_k^c \wedge x_{k-1} \in \mathbb{X}_{k-1}^c\right\},$$

and corresponds to the set of inputs that produces the set of feasible states.

Remark 8.1. Note that $\mathbb{U}_{k-1}^c$ in Definition 8.6 is an alternative way of expressing the admissible input sequence in (2.3). ◊

8.3 Admissibility Evaluation Approaches

8.3.1 Admissibility Evaluation Using Constraint Satisfaction

The admissibility evaluation problem for a given AFC can be handled naturally as a CSP. This section deals with the proposed methodology for evaluating the admissibility of a particular AFC by means of the constraint satisfaction approach. First, some CSP-related concepts are presented and particular details related to this approach are discussed. This approach is then explained in the framework of the AFC admissibility evaluation.

The Concept of CSP

A CSP for a set can be formulated as a 3-tuple $\mathscr{H} = (\Upsilon, \Lambda, \mathbb{C})$ (Jaulin *et al.*, 2001a), where:

- $\Upsilon = \{\upsilon_1, \cdots, \upsilon_n\}$ is a finite set of n variables
- $\Lambda = \{\Lambda_1, \cdots, \Lambda_n\}$ is the set of their domains represented by closed sets
- $\mathbb{C} = \{c_1, \cdots, c_n\}$ is a finite set of constraints relating to the variables in Υ.

A point solution of $\mathscr{H}$ is an n-tuple $(\tilde{\upsilon}_1, \cdots, \tilde{\upsilon}_n) \in \Lambda$ such that all of the constraints $\mathbb{C}$ are satisfied. The set of all solution points of $\mathscr{H}$ is denoted by $\mathbb{S}(\mathscr{H})$. This set is called the *global solution set*. The variables $\upsilon_i \in \Upsilon_i$ are *consistent* in $\mathscr{H}$ if and only if

$$\forall \upsilon_i \in \Upsilon_i \; \exists \, (\tilde{\upsilon}_1 \in \Lambda_1 \cdots, \tilde{\upsilon}_n \in \Lambda_n) \mid (\tilde{\upsilon}_1, \cdots, \tilde{\upsilon}_n) \in \mathbb{S}(\mathscr{H}),$$

with $i = 1, \dots, n$. The solution of a CSP is said to be *globally consistent* if and only if every variable is consistent. A variable is *locally consistent* if and only if it is consistent with respect to all directly connected constraints. Thus, the solution of a CSP is said to be locally consistent if all variables are locally consistent.

The principle of an algorithm that solves a CSP using local consistency techniques is essentially to iterate two main operations, *domain contraction* and *propagation*, until a stable state is reached. Roughly speaking, if the domain of a variable v_i is locally contracted with respect to a constraint c_j, then this domain modification is propagated to all of the constraints in which v_i occurs, leading to the contraction of other variable domains, and so on. Thus, the final goal of such a strategy is to contract the domains of the variables as much as possible without losing any solution by removing inconsistent values through the *projection* of all constraints. Projecting a constraint with respect to some of its variables means computing the smallest set that contains only consistent values by applying a contraction operator.

Since they are incomplete by nature, these methods must be combined with enumeration techniques (*e.g.*, bisection) in order to distinguish the solutions whenever possible. Domain contraction utilises *contraction operators* that compute over approximate real-number domains (Jaulin *et al.*, 2001b).

Admissibility Evaluation Approach

The admissibility evaluation of a given AFC requires the computation of the admissible solutions set introduced in Definition 8.3. Note that this naturally corresponds to a CSP for sets. Algorithm 8.1 allows the admissibility of a given AFC to be evaluated by solving the CSP defined by the dynamical model of the system, the operative limits on inputs and states over H_p, and the initial states.

It is well known that solving these problems involves high computational complexity, as to represent the solution sets accurately they need to be decomposed using *subpavings*, which implies exponential computation times (DiLoreto *et al.*, 2007). One relaxation that can be utilised to reduce the computational complexity associated with Algorithm 8.1 is to approximate the variable domain sets with their interval hulls. Hence, the new set of domains for $\mathscr{H} = (\Upsilon, \Lambda, \mathbb{C})$ is expressed as

$$\Lambda = \left\{ \square\mathbb{X}_1, \square\mathbb{X}_2,, \cdots, \square\mathbb{X}_{H_p}, \square\mathbb{U}_0, \square\mathbb{U}_2, \cdots, \square\mathbb{U}_{H_p-1}, \square\mathbb{J}_{\mathbb{A}} \right\}.$$

This procedure avoids the need for set computations based on a subpaving approach and allows the use of interval methods that have efficient operators (contractors) for identifying the solution of an *interval* CSP (ICSP) (see Hyvnen, 1992). However, the interval hull determination of the sets that define the variable domains (*i.e.*, the interval box that approximates these

Algorithm 8.1 Admissibility evaluation using CSP for sets

1: **for** $k = 1$ to H_p **do**
2: $\quad \mathbb{U}_{k-1} \Leftarrow \mathbb{U}$
3: $\quad \mathbb{X}_k \Leftarrow \mathbb{X}$
4: **end for**
5: $\Upsilon \Leftarrow \{\overbrace{x_{1|k}, x_{2|k}, \cdots, x_{H_p|k}}^{\mathbf{x}_k}, \overbrace{u_{0|k}, u_{1|k}, \cdots, u_{H_p-1|k}}^{\mathbf{u}_k}, J\}$
6: $\Lambda \Leftarrow \{\mathbb{X}_1, \mathbb{X}_2, \cdots, \mathbb{X}_{H_p}, \mathbb{U}_0, \mathbb{U}_1, \cdots, \mathbb{U}_{H_p-1}, \mathbb{J}_{\mathbb{A}}\}$
7: $\mathbb{C} \Leftarrow \left\{ \mathbf{x}_{k+1} = A\mathbf{x}_k + B\mathbf{u}_k, J(\mathbf{x}_k, \mathbf{u}_k) = \phi(x_{H_p|k}) + \sum_{i=0}^{H_p-1} \Phi(x_i, u_i) \right\}$
8: $\mathscr{H}_{\mathbb{A}} = (\Upsilon, \Lambda, \mathbb{C})$
9: $\mathbb{A} = \mathbf{solve}(\mathscr{H}_{\mathbb{A}})$
10: **if** $\mathbb{A} = \emptyset$ **then**
11: $\quad$ AFC is *inadmissible*
12: **else**
13: $\quad$ AFC is *admissible*
14: **end if**

sets more closely) requires global consistency, which still comes at a high computational cost (Hyvnen, 1992). To reduce this computational complexity, an ICSP can be solved by using contractors, although this only guarantees local consistency; *i.e.*, it is only guaranteed that every constraint in the ICSP is consistent independent of any other constraint (Jaulin *et al.*, 2001a). The use of local consistency implies that the intervals that restrict the variable domain sets are not the most well adjusted, but the intervals obtained must still contain the solution sets. To obtain the most well-adjusted intervals, global consistency would be required (*i.e.*, it would be necessary to guarantee that all constraints in the ICSP were consistent simultaneously).

An alternative approach to solving the CSP proposed in Algorithm 8.1 is to allow the relations between variables to be ruptured for consecutive time instants, which makes it possible to determine the interval hull for the set of feasible solutions step by step. However, the problem of uncertainty propagation (the wrapping effect) could appear when the CSP is solved in this way, since an interval simulation problem is also being implicitly solved. This problem does not appear in *isotonic systems*[2] (Cugueró *et al.*, 2002), which are systems in which states and input variables g are mapped isotonically.[3] In this case, it is only necessary to propagate the interval hull of the admissible solutions set one step ahead of the current iteration.

To summarise, Algorithm 8.1 can be approximated as presented in Algorithm 8.2 by:

- Relaxing a CSP for sets to a CSP over intervals
- Using local consistency instead of global consistency, and

[2] See also *monotonic systems* (Angeli and Sontag, 2003).

[3] A generic function $g = (g_1, g_2, \ldots, g_n)$ is isotonic about x if none of the g_i decrease with respect to all of the x_j: $j = 1, \ldots, n$.

Algorithm 8.2 Admissibility (iterative) evaluation using interval CSP (ICSP)

$\square\mathbb{X}_0 \Leftarrow \square\mathbb{X}$
$\square\mathbb{J}_0 \Leftarrow \square\mathbb{J}_{\mathbb{A}_0}$
for $k = 1$ to H_p **do**
    $\square\mathbb{X}_k \Leftarrow \square\mathbb{X}$
    $\square\mathbb{U}_{k-1} \Leftarrow \square\mathbb{U}$
    $\Upsilon \Leftarrow \{x_k, x_{k-1}, u_k, J_k, J_{k-1}\}$
    $\Lambda \Leftarrow \{\square\mathbb{X}_k, \square\mathbb{X}_{k-1}, \square\mathbb{U}_{k-1}, \square\mathbb{J}_{\mathbb{A}k}, \square\mathbb{J}_{k-1}\}$
    $\mathbb{C} \Leftarrow \{x_k = Ax_{k-1} + Bu_{k-1}, J_k = J_{k-1} + \Phi(x_{k-1}, u_{k-1})\}$
    $\mathscr{H}_{\square\mathbb{A}_k} = (\Upsilon, \Lambda, \mathbb{C})$
    $(\square\mathbb{A}_k, \square\mathbb{X}_k, \square\mathbb{U}_{k-1}, \square\mathbb{J}_k) = \textbf{solve}(\mathscr{H}_{\square\mathbb{A}_k})$
end for
$\square\mathbb{A} = \bigcup_{k=0}^{H_p} \square\mathbb{A}_k$
if $\square\mathbb{A} = \emptyset$ **then**
    AFC is *inadmissible*
else
    Admissibility of the AFC is *undefined*
end if

- Computing the domains of the CSP variables iteratively from the previously computed domains.

Algorithm 8.2 enables the admissibility evaluation of an AFC given the interval hull of a particular admissible control objective set $\mathbb{J}_{\mathbb{A}}$. Several algorithms can be used to solve the ICSP articulated in this algorithm, including Waltz's local filtering algorithm (Waltz, 1975) and Hyvnen's tolerance propagation algorithm (Hyvnen, 1992). The former only ensures locally consistent solutions while the latter can guarantee globally consistent solutions, although with high computational complexity.

Remark 8.2. If the interval hull of the admissible solution set $\mathbb{A}$ returned by Algorithm 8.2 is empty, then $\mathbb{A}$ is an empty set too, and the AFC is therefore inadmissible. Otherwise, nothing can be stated, as $\square\mathbb{A} \neq \emptyset$ does not imply $\mathbb{A} = \emptyset$. ◊

An Illustrative Example

The presence of constraints in MPC makes it very difficult to proceed with the fault-tolerance analysis proposed by Staroswiecki (2003), although analytical results could be obtained in that work as the way that the fault affected the objective function could be expressed using LQR theory. Nevertheless, this expression is not available in constrained MPC, even if it is possible to derive an explicit expression for the controller (Bemporad *et al.*, 2002b). This fact highlights the utility of the proposed approach.

The double integrator system proposed by Bemporad *et al.* (2002b) is considered here. The equivalent discrete-time state-space description of this using the Euler discretisation rule is

$$x_{k+1} = \begin{bmatrix} 1 & 1 \\ 0 & 1 \end{bmatrix} x_k + \begin{bmatrix} 0 \\ 1 \end{bmatrix} u_k \tag{8.3}$$

$$y_k = \begin{bmatrix} 1 & 0 \end{bmatrix} x_k, \tag{8.4}$$

where the system states and the control signal are constrained as $x_1 \in [-15,15]$, $x_2 \in [-6,6]$ and $u \in [-1,1]$. An MPC controller is employed to control this system while satisfying the associated state and control constraints and considering the following objective function:

$$J = x_{H_p}^T P x_{H_p} + \sum_{i=0}^{H_p-1} \left(x_i^T Q x_i + R u_i^2\right), \tag{8.5}$$

where $H_p = 2$ and the terminal weight matrix P is determined using the DARE with $Q = \text{diag}([1 \quad 0])$ and $R = 0.01$. According to Bemporad *et al.* (2002b), a state-feedback explicit control law of the form

$$u_k = K_{\text{PWA}}(x_k) x_k, \tag{8.6}$$

which is piecewise affine with respect to the states, can be derived. Using the Hybrid Toolbox for MATLAB® (Bemporad, 2006), the expression for K_{PWA} can be determined and represented graphically (see Figure 8.1). The closed-loop state trajectory can be computed and represented using this law.

Using the CSP-based method for AFC admissibility evaluation proposed in this chapter, the feasible sets for states x_1 and x_2 were computed and are shown in Figures 8.2 (a) and (b), respectively. Note that the closed-loop state trajectories upon applying the MPC controller are inside the corresponding feasible sets, as expected.

Now, suppose that a fault in the actuator is introduced. This fault reduces the operating range of the actuator such that $u \in [-0.75, 0.75]$. When the feasible sets for states x_1 and x_2 are recomputed, it is apparent that the closed-loop state trajectories upon applying the MPC controller to the faulty situation are outside the corresponding feasible sets for the faulty situation. This means that the performance of the MPC controller is worse than it was in the case of the fault-free actuator, as the MPC trajectories for the faulty situation are not reachable. If the MPC trajectories in the faulty situation were inside the corresponding feasible sets, the performance of the MPC controller would not be affected by the fault; *i.e.*, it would be fault tolerant. This example shows how easily the tolerance of a control law with respect to a fault can be evaluated using the method proposed.

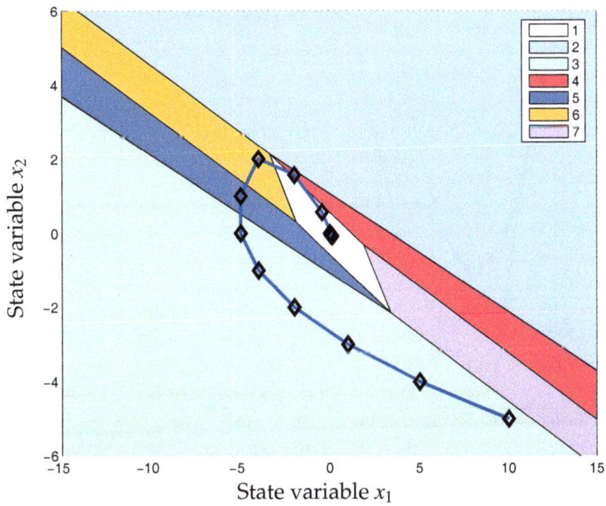

Figure 8.1 Explicit state-space polyhedral partition and MPC law for the illustrative example

8.3.2 *Admissibility Evaluation Using Set Computation*

The iterative algorithm (8.2) cannot be applied to non-isotonic systems. To extend the applicability of this algorithm to non-isotonic systems, the following alternatives may be considered:

- Approximate the feasible solution domains with more complex domain forms than an interval hull, such as zonotopes, ellipses, *etc.*, and use set propagation and/or set constraint satisfaction
- Convert the system into an isotonic system via state feedback techniques (Maciejowski, 2002)
- Formulate a CSP by propagating the initial state and using global consistency techniques
- Formulate the problem in an analytical way (linear system) and use the corresponding tools to find the solution (Bemporad *et al.*, 2002b).

This section discusses the application of the first option using the associated operations for set computation. Moreover, step 12 of Algorithm 8.2 (the solution of the CSP) is replaced by a set propagation algorithm that requires the computation of all of the sets defined in Section 8.2. The evaluation of the admissibility of a given AFC using the set computation approach presented in Algorithm 8.3 follows the iterative method proposed in Algorithm 8.2. This section does not discuss how the set computations involved are per-

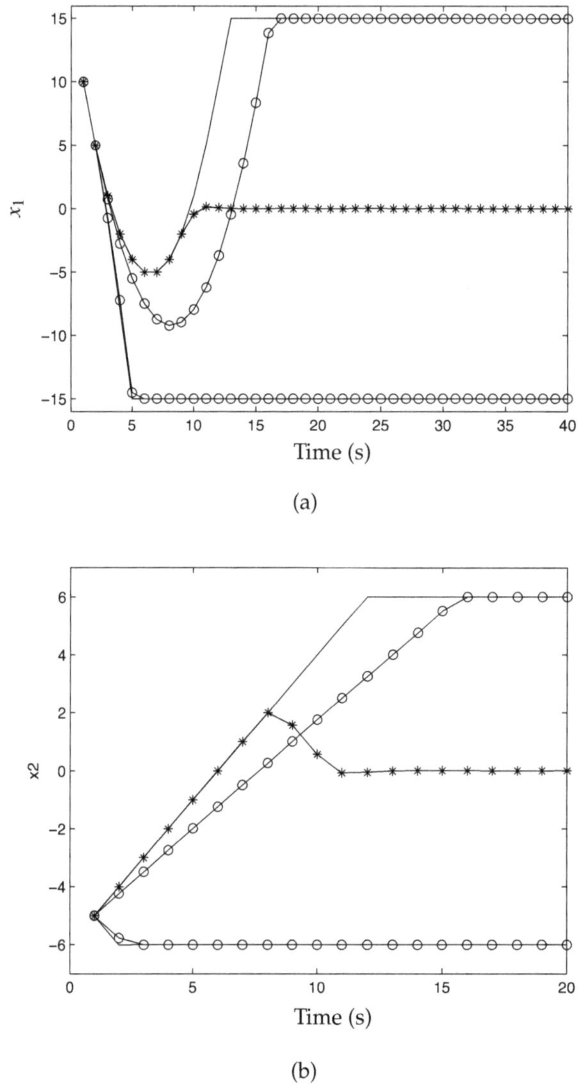

Figure 8.2 Feasible sets corresponding to the state variables in fault-free (*solid lines with no symbols*) and faulty (*lines with circles*) scenarios for the illustrative example. Feasible state trajectories for (a) x_1, and (b) x_2 are shown. The MPC solution for a faulty situation is also presented (*lines with asterisks*)

formed. In any case, set computations based on zonotopes and related set operations are described in the next section.

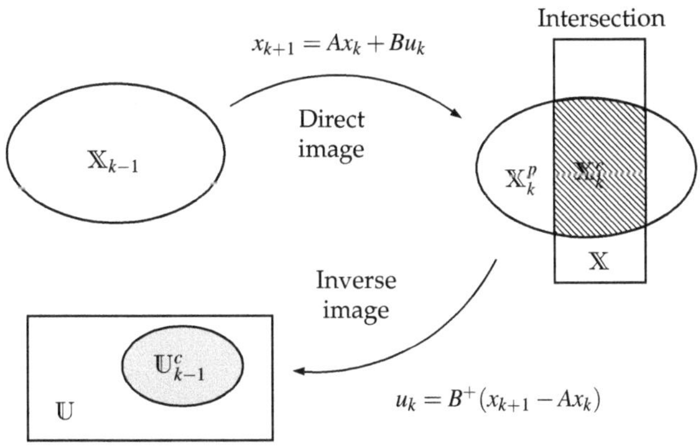

Figure 8.3 Graphical interpretation of Algorithm 8.1

Algorithm 8.3 Computation of Ω

1: $\mathbb{X}_0 \Leftarrow \mathbb{X}$
2: $\Omega_0 \Leftarrow \mathbb{X}_0 \times \mathbb{U}$
3: **for** $k = 1$ to H_p **do**
4: $\quad \mathbb{U}_{k-1} \Leftarrow \mathbb{U}$
5: $\quad$ Compute $\mathbb{X}_k^p$ from $\mathbb{X}_{k-1}$ and $\mathbb{U}_{k-1}$ % *see Definition 8.4*
6: $\quad$ Compute $\mathbb{X}_k^c = \mathbb{X} \cap \mathbb{X}_k^p$ % *see Definition 8.5*
7: $\quad$ Compute $\mathbb{U}_{k-1}^c$ from $\mathbb{X}_k^c$ % *see Definition 8.6*
8: $\quad \Omega_k = \mathbb{X}_k^c \times \mathbb{U}_{k-1}^c$
9: $\quad \mathbb{X}_k \Leftarrow \mathbb{X}_k^c$
10: **end for**
11: $\Omega = \bigcup_{k=0}^{H_p} \Omega_k$

Admissibility Evaluation Approach

Algorithm 8.3 starts by obtaining the feasible solutions set Ω given a set of initial states $\mathbb{X}_0 \subseteq \mathbb{X}$, the system equations and the system operating constraints over H_p. At each iteration, the predicted states set $\mathbb{X}_k$ (see *Definition* 8.4) is computed (step 5). The computation of this set requires the computation of the direct image of the system state function (2.1) (see Figure 8.3). Such a set is obtained by considering that the whole operational range of the system actuators is available (step 4). The set of compatible states in Definition 8.5 is then determined by considering the constraints on the states (step 6) using an intersection operation. This in turn allows to determine the compatible input set in Definition 8.6 (step 7) using the inverse image of the system state function (2.1) (see Figure 8.3).

When the set Ω is being computed, the set of feasible control objectives (Definition 8.2) can be obtained at the same time as the direct image of the

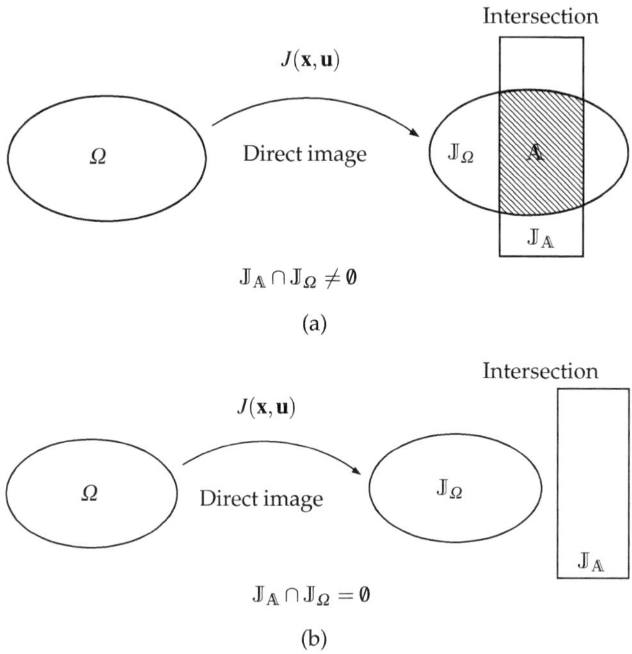

Figure 8.4 Graphical interpretation of Algorithm 8.2: (a) admissible AFC, and (b) inadmissible AFC

Algorithm 8.4 Admissibility evaluation of AFC given an admissible set of control objectives $\mathbb{J}_{\mathbb{A}}$

1: $\mathbb{X}_k \Leftarrow \mathbb{X}_0$
2: $\Omega_0 \Leftarrow \mathbb{X}_0$
3: **for** $k = 1$ to H_p **do**
4:     Compute Ω_k using Algorithm 8.3
5:     Compute $\mathbb{J}_{\Omega_k}$ *% see Definition 8.2*
6: **end for**
7: $\mathbb{J}_\Omega = \bigcup_{k=0}^{H_p} \mathbb{J}_{\Omega_k}$
8: **if** $\mathbb{J}_{\mathbb{A}} \cap \mathbb{J}_\Omega = \emptyset$ **then**
9:     AFC is *inadmissible*
10: **else**
11:     AFC is *admissible*
12: **end if**

---

control objective function 8.2 (see Figure 8.4). Thus, at time $k = H_p$, the set $\mathbb{J}_{\Omega_k}$ is obtained by following the procedure outlined in Algorithm 8.4.

Once $\mathbb{J}_{\Omega_f}$ has been obtained from step 7 of Algorithm 8.4, step 8 yields the result of the AFC admissibility evaluation when a set of admissible control objectives $\mathbb{J}_{\mathbb{A}}$ is provided. More precisely, the AFC is inadmissible if

$$\mathbb{J}_{\Omega_f} \cap \mathbb{J}_{\mathbb{A}} = \emptyset. \tag{8.7}$$

In this case, the admissible solutions set $\mathbb{A}$ is empty. Otherwise, the AFC is admissible and so the admissible solutions set $\mathbb{A}$ is not empty (see Figure 8.4).

Set Computations Using Zonotopes

Except in very particular cases, it is not possible to evaluate the sets required in Algorithm 8.3 exactly. Instead, guaranteed outer approximations (as accurate as possible) of these sets can be used, as discussed previously. In this section, zonotopes and related operations are used to compute these outer approximations. First of all, the following definitions are supplied:

Definition 8.7 (Minkowski Sum). The Minkowski sum of two sets $\mathbb{X}$ and $\mathbb{Y}$ in a Euclidean space is defined as

$$\mathbb{X} \oplus \mathbb{Y} = \{x + y \,:\, x \in \mathbb{X}, y \in \mathbb{Y}\}.$$

This operation is also known as *dilatation.*

Definition 8.8 (Zonotope). Given a vector $p \in \mathbb{R}^n$ and a matrix $H \in \mathbb{R}^{n \times m}$, the set represented as

$$\mathbb{X} = p \oplus HB^m = \{p + Hz \,:\, z \in B^m\}$$

is called a *zonotope* of order m, and this corresponds to the Minkowski sum of the segments defined by the columns of matrix H. In this expression, B^m is an unitary box consisting of m unitary intervals.

Note that the order of the parameter m can be considered a measure of the geometrical complexity of the zonotope. For instance, Figure 8.5 shows the construction of a zonotope from a sequence of values of m. As m increases, the zonotope topology becomes more complex.

Computation of the predicted states set $\mathbb{X}_k^p$ (step 5 of Algorithm 8.3). The computation of the prediction set $\mathbb{X}_k^p$ can be viewed as the evaluation of the *direct image* of (2.1) using $A\,\mathbb{X}_k^c + B\,\mathbb{U}_k^c$. This image can be bounded using zonotopes as follows. Given two zonotopes $\mathbb{X}_k^c = p_{x_k} \oplus H_{x_k} B^{r_1}$ and $\mathbb{U}_k^c = p_{u_k} \oplus H_{u_k} B^{r_2}$, then

$$\mathscr{X}_{k+1}^p = p_{k+1} \oplus H_{k+1} B^{r_n}, \tag{8.8}$$

where

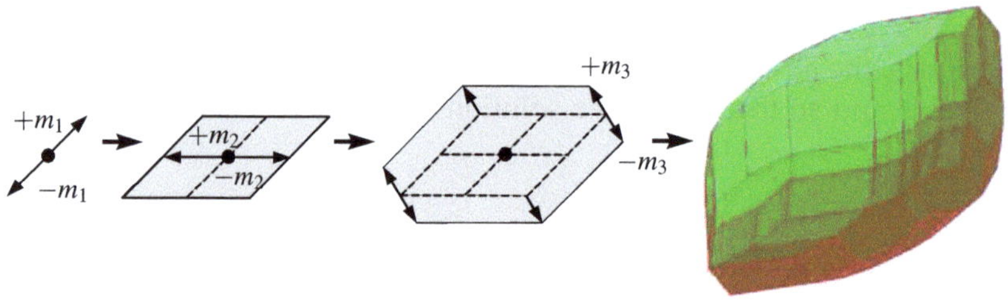

Figure 8.5 Zonotope construction. Sequence from $m=1$, $m=2$, $m=3$ and $m=15$

$$p_{k+1} = Ap_{x_k} + Bp_{u_k}, \tag{8.9}$$
$$H_{k+1} = \begin{bmatrix} AH_{x_k} & BH_{u_k} \end{bmatrix}. \tag{8.10}$$

It is important to note that this method of computing the set of predicted states increases the number of segments when the zonotope $\mathbb{X}^p_{k+1}$ is generated. Therefore, in order to control the domain complexity, a reduction step is implemented. The method proposed in Combastel (2003) for reducing the zonotope complexity is employed. This is described in Property 8.1.

Property 8.1. Given the zonotope $\mathbb{X} = p \oplus HB^r \subset \mathbb{R}^n$ and the integer s (with $n < s < r$), let $\widehat{H}$ be the matrix that results when the H columns are reordered according to decreasing Euclidean norm. Then

$$\mathbb{X} \subseteq p \oplus \begin{bmatrix} \widehat{H}_T & Q \end{bmatrix} B^s, \tag{8.11}$$

where $\widehat{H}_T$ is obtained from the first $s-n$ columns of matrix $\widehat{H}$, and $Q \in \mathbb{R}^{n\times n}$ is a diagonal matrix that satisfies

$$Q_{i,i} = \sum_{j=s-n+1}^{r} |\widehat{H}_{ij}|, \qquad i = 1, \cdots, n. \tag{8.12}$$

Computation of $\mathbb{X}^c_k$ (step 6 of Algorithm 8.3). In the intersection set step, it is necessary to characterise the set $\mathbb{X}^c_k$. This set corresponds to the intersection of two sets: the previously bounded set $\mathbb{X}^p_k$ and the set that restricts the states for all k over H_p. Property 8.2 defines the intersection of two zonotopes.

Property 8.2. Given two zonotopes $\mathbb{X}_1 = p_1 \oplus H_1 B^{r_1}$ and $\mathbb{X}_2 = p_2 \oplus H_2 B^{r_2}$, and a matrix E of appropriate dimensions, the following expressions can be stated:

$$\widehat{p}(E) = Ep_1 + (I - E)p_2, \tag{8.13}$$

$$\widehat{H}(E) = [EH_1 \ (I - E)H_2]. \tag{8.14}$$

Therefore, the intersection of the zonotopes $\mathbb{X}_1$ and $\mathbb{X}_2$ is given by

$$\mathbb{X}_1 \cap \mathbb{X}_2 \subseteq \widehat{\mathbb{X}}(E), \tag{8.15}$$

$$\widehat{\mathbb{X}}(E) = \widehat{p}(E) \oplus \widehat{H}(E) B^{r_1 + r_2}. \tag{8.16}$$

In order to reduce the size of the intersection zonotope $\widehat{\mathbb{X}}(E)$, a convex optimisation problem has to be solved. If H_{1i} and H_{2j} (with $i = 1, \cdots, m_1$, $j = 1, \cdots, m_2$) are the columns of matrices H_1 and H_2, then the function to be minimised is

$$f(E) = \sum_{i=1}^{m_1} (EH_{1i})^T (EH_{1i}) + \sum_{j=1}^{m_2} (H_{2j} - EH_{2j})^T (H_{2j} - EH_{2j}). \tag{8.17}$$

Computation of $\mathbb{U}^c_{k-1}$ (step 7 of Algorithm 8.3). Before the computation of this set is presented, Definition 8.9 is introduced.

Definition 8.9 (Minkowski Difference). The *Minkowski* (or *Pontryagin) difference* between two given sets $\mathbb{X}, \mathbb{Y} \subset \mathbb{R}^n$ is defined as

$$\mathbb{X} \sim \mathbb{Y} \triangleq \{x \in \mathbb{R}^n \ : \ x + y \in \mathbb{X}, \ \forall y \in \mathbb{Y}\}.$$

The Minkowski difference is a special case of the *Minkowski sum* (Definition 8.7) of two convex shapes.

Thus, the set $\mathbb{U}^c_{k-1}$ can be computed from the Minkowski difference between the zonotopes $\mathbb{X}^c_k$ and $\mathbb{X}^c_{k-1}$, considering that the system matrix B has full row rank (Mayne and Schroeder, 1997). This difference can be expressed as

$$\mathbb{U}^c_{k-1} = B^+ (\mathbb{X}^c_k \sim A\mathbb{X}^c_{k-1}), \tag{8.18}$$

where B^+ denotes the *pseudoinverse* of matrix B. In the case that matrix B does not have full row rank, *singular value decomposition* (SVD) should be applied to it in order to compute the matrix B^+ (Theilliol *et al.*, 2002).

The difference between zonotopes $\mathbb{X}_k^c$ and $\mathbb{X}_{k-1}^c$ is computed by applying interval arithmetic (Moore, 1966), which employs the interval hulls of both zonotopes. Notice that the definition of the interval hull can be applied to zonotopes in a straightforward manner. Given the zonotope $\mathbb{X} = p \oplus HB^r$, its interval hull is computed as

$$\Box\mathbb{X} = p \oplus rs(H)B^r, \tag{8.19}$$

where

$$rs(H)_{ii} = \sum_{j=1}^{p} |H_{ij}|. \tag{8.20}$$

Here, rs is the *row sum* of matrix H, which results in a diagonal matrix of suitable dimensions.

8.4 Actuator Fault Tolerance Evaluation in Sewer Networks

8.4.1 System Description

In order to show how to apply the AFC admissibility evaluation to sewer networks, a small system inspired by the BTC will be employed. This three-tank catchment (3-TC) contains the representative elements of the entire sewer network and considers three of the four control gates that appear in the BTC. This 3-TC system is therefore sufficient to illustrate the effectiveness of the admissibility evaluation methods discussed in this chapter.

The 3-TC, which is presented in Figure 8.6, is described by the discrete-time expression in (4.1), where

$$A = \begin{bmatrix} 1-\Delta t\beta_2 & 0 & 0 \\ 0 & 1 & 0 \\ \Delta t\beta_2 & 0 & 1-\Delta t\beta_4 \end{bmatrix}, \quad B = \Delta t \begin{bmatrix} 1 & 0 & 0 \\ 0 & 1 & -1 \\ -1 & -1 & 1 \end{bmatrix}, \quad B_p = \Delta t \begin{bmatrix} 0 & \alpha_2 & 0 \\ 0 & 0 & 0 \\ 1 & 0 & \alpha_4 \end{bmatrix}.$$

In Figure 8.6, d_1 corresponds to a rain inflow. This is because the virtual tank T_1 is not considered, as no gate has any influence on its dynamical behaviour. However, $d_{2k} = \alpha_2 P_{16k}$ and $d_{4k} = \alpha_4 P_{20k}$ are the products of measurements from the rain gauges (P_i) and the conversion coefficients $\alpha_2 = \varphi_2 S_2 = 0.5951$ and $\alpha_4 = \varphi_4 S_4 = 0.1530$. Parameters β_i, φ_i and S_i, for $i \in \{2,3,4\}$, are taken from Table 3.1 in Chapter 3.

The system constraints expressed in the notation adopted in this book are written as:

- Bounding constraints, which refer to physical restrictions of the system variables:

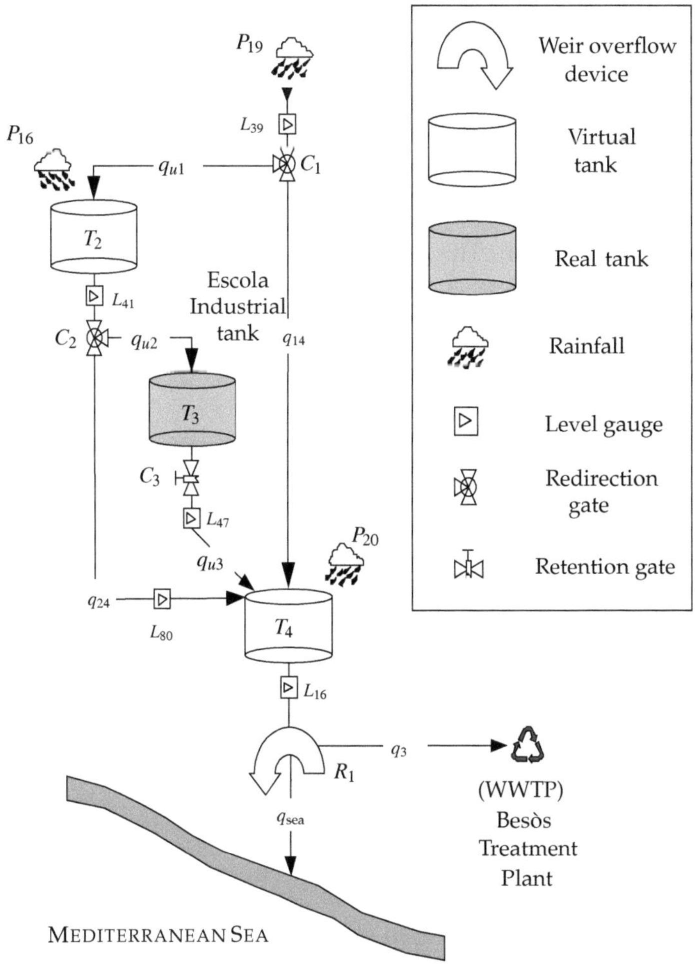

Figure 8.6 Three-tank catchment (3-TC)

$$
\begin{aligned}
v_{2k} &\in [0, +\infty], & q_{u_1 k} &\in [0, 11],\\
v_{3k} &\in [0, 35000], & q_{u_2 k} &\in [0, 25],\\
v_{4k} &\in [0, +\infty], & q_{u_3 k} &\in [0, 7]
\end{aligned}
\tag{8.21}
$$

- Mass conservation constraints:

$$
\begin{aligned}
d_{1k} &= q_{u_1 k} + q_{14k},\\
q_{x_1 k} &= q_{u_2 k} + q_{24k},\\
q_{x_2 k} &\geq q_{u_3 k}.
\end{aligned}
\tag{8.22}
$$

For this application, it is assumed that vector d_k (rain) is known at each time instant $k \in \mathbb{Z}_+$, which implies a known perturbation. It is desired to evaluate the admissibilities of different AFCs in both reconfiguration and accommodation schemes. The admissibility of a given AFC is evaluated according to the amount of degradation in the control objectives compared to their values under nominal (fault-free) conditions during a given rain episode. The selected rainstorm is the one that occurred on September 14, 1999; see Figure 3.7 (a). On that day, severe flooding occurred as a consequence of the rainstorm. The actuator faults do not occur simultaneously, and they are modelled as changes in operating limits or complete losses (cancellation) of actuator function.

Setting Up Algorithm 8.2

At each time instant k, the ICSP $\mathscr{H} = (\Upsilon, \Lambda, \mathbb{C})$ associated with the system has a set of variables with nine components:

$$\Upsilon = \left\{v_{2k}, v_{3k}, v_{4k}, v_{2k+1}, v_{3k+1}, v_{4k+1}, q_{u_1 k}, q_{u_2 k}, q_{u_3 k}\right\},$$

as well as the domains set

$$\Lambda = \{[v_2]_k, [v_3]_k, [v_4]_k, [0, +\infty], [0, 35000], [0, +\infty], [0, 11], [0, 25], [0, 7]\},$$

and the set of constraints given by the corresponding system model (4.1) and the expressions in (8.21) and (8.22).

Solving the ICSP

Several algorithms can be used to solve the ICSP articulated in Algorithm 8.2, including Waltz's local filtering algorithm (Waltz, 1975) and Hyvnen's tolerance propagation algorithm (Hyvnen, 1992). The former only ensures locally consistent solutions, while the latter can guarantee globally consistent solutions. In this chapter, the ICSP is solved using a tool based on interval constraint propagation known as *Interval Peeler* (Baguenard, 2005). The goal of this software is to determine the solution to the ICSP defined in Section 8.3.1 in the case where the domains are represented by closed real intervals. The solution provides refined interval domains that are consistent with the set of ICSP constraints. Other solvers, such as *RealPaver* (Granvilliers and Benhamou, 2006), perform ICSP computations using Cartesian products of intervals. In this solver, several consistency techniques (box, hull, and 3B) are implemented.

Remark 8.3. In order to evaluate the fault tolerance of an LCMPC controller after the occurrence of a fault, it is not necessary to find a point solution to the CSP, only whether the CSP has a solution. In particular, the fact that a CSP has no solution means that the LCMPC controller is not fault tolerant with respect to the considered AFC.

The existence of a solution to the CSP associated with the tolerance evaluation of the LCMPC controller for a given AFC can be checked for by solving the following OOP:

$$\min_{\{u \in \mathbb{R}^m\}} h(\cdot), \tag{8.23a}$$

subject to

$$J(\mathbf{x}_k, \mathbf{u}_k) \leq J_f, \tag{8.23b}$$

$$x_{0|k} = x_k, \tag{8.23c}$$

$$\mathbf{x}_{k+1} = g(\mathbf{x}_k, u_i) \tag{8.23d}$$

for $i = \{0, 1, \ldots, H_p - 1$, where $J(\mathbf{x}_k, \mathbf{u}_k)$ is from (8.2) and $g(\mathbf{x}_k, u_i)$ is a linear mapping of the states. Here, $h(\cdot)$ denotes the null function, since it is only important to know whether or not the problem constraints are violated. If the problem is feasible, the CSP has a solution and the LCMPC controller is fault tolerant with respect to the control objective (8.2) with a acceptable degradation that is less than or equal to J_f, which establishes the admissibility threshold. Also note that this evaluation can be applied whether the system prediction model is of a hybrid or nonlinear nature, since the OOP statement follows the same philosophy. ◊

8.4.2 Control Objectives and Admissibility Criterion

The main control objectives are defined as the minimisation of the pollution and the minimisation of the CSO into streets (flooding caused by the insufficient capacities of sewer mains q_{14} and q_{24}). Note that overflows from virtual tanks are not considered in this case. From the system variables, the expressions that define the control objectives are as follows:

- For pollution:

$$V_{\text{pll}} = \Delta t \sum_{k=0}^{H_p} \max(0, q_{v_4 k} - \overline{q}_4) \tag{8.24}$$

- For flooding:

$$V_{\text{fld}} = \Delta t \sum_{k=0}^{H_p} \max(0, q_{14k} - \overline{q}_{14}) + \max(0, q_{24k} - \overline{q}_{24}). \tag{8.25}$$

Also note that the pollution is expressed as a function of an isotonic state variable, for which an interval hull can be computed exactly. Therefore, since the pollution depends only on this variable, its interval hull is also exact. In this case, $\square\mathbb{J}_{\square\Omega} \supseteq \square\mathbb{J}_{\mathbb{A}}$ holds with equality, which means that the admissibility evaluation is always correct.

On the other hand, when the control objective related to the flooding is taken into account, it can be seen that its expression depends on relations between isotonic and nonisotonic variables. Consequently, from Remark 8.2, an assessment of the inadmissible configuration is possible, but nothing can be stated about the admissibility of that AFC.

In particular, the admissibility criterion is based on a comparison between the resulting minimum volumes (of either pollution or flooding) at the end of the scenario for the nominal and faulty cases. Denoting $\chi_{\text{pll}}(H_p)$ and $\chi_{\text{fld}}(H_p)$ as the minimum pollution and flooding volumes at time $k = H_p$, respectively, the comparison described above is given by

$$\chi^{\text{f}}_{\text{pll}\,H_p} = \psi\,\chi^{\text{nom}}_{\text{pll}\quad H_p}, \tag{8.26a}$$

$$\chi^{\text{f}}_{\text{fld}\,H_p} = \psi\,\chi^{\text{nom}}_{\text{fld}\quad H_p}, \tag{8.26b}$$

where ψ is related to the degradation and superscripts f and nom are related to the system status (*i.e.*, nominal or faulty cases). In order to illustrate the proposed evaluation methodology for this application, it is assumed that $\psi = 8$. Actually, this relation is provided by the network operator according to the directives defined by the city authorities, which in turn are based on the heuristic knowledge of the sewage system designers.

8.4.3 Main Results

Actuator Cancellation

This case considers actuators that are completely closed or completely open due to the occurrence of a fault, which changes the admissibility of the resultant AFC. Table 8.1 summarises the possible fault cases and their admissibilities when pollution is considered. Only faults where the actuators were completely closed are simulated; *i.e.*, $q_{u_i} \in [0,0]$ and $q_i \in [0,+\infty]$. By contrast, the ICSP cannot be solved because the bounding constraints of q_{u_i} are violated.

On the other hand, Table 8.2 presents the results obtained when flooding is considered. Note that some AFC evaluations are uncertain due to the definition of the cost function for this objective (see Section 8.4.2).

Table 8.1 Actuator cancellation: admissibility of AFCs when pollution is considered

Fault location	$\chi_{\text{pll}_{H_p}}$ (m^3)	Admissibility evaluation
No fault	1050	—
Fault in q_{u_1}	8800	Inadmissible
Fault in q_{u_2}	52200	Inadmissible
Fault in q_{u_3}	1050	Admissible

Table 8.2 Actuator cancellation: admissibility of AFCs when flooding is considered

Fault location	$\chi_{\text{fld}_{H_p}}$ (m^3)	Admissibility evaluation
No fault	5100	—
Fault in q_{u_1}	5100	Uncertain
Fault in q_{u_2}	73200	Inadmissible
Fault in q_{u_3}	5100	Uncertain

Figure 8.7 shows the minimum values for pollution (Figure 8.7 (a)) and for flooding (Figure 8.7 (b)) when the given admissibility criterion is considered (threshold).

Partially Damaged Actuators

This case considers that faults manifest themselves as reduced operating ranges (*e.g.*, 0–100% to 0–50%). Thus, the number of admissible AFCs varies as shown in Tables 8.3 and 8.4; if the admissibility criterion in (8.26) is maintained, then more AFCs are admissible. Different actuator ranges are presented. Note that the data in these tables were obtained by assuming that there is no fault at q_{u_3} due to the insensitivity of this actuator to the system, as seen in the results collected in Tables 8.1 and 8.2. Figure 8.8 presents the minimum volumes of pollution and flooding observed in this faulty actuator case.

8.5 Summary

This chapter proposed a method for evaluating the admissibility of AFCs that was based on solving an interval CSP (ICSP). This procedure implies the propagation of the feasible solution set computed at each time instant by implicitly solving an interval simulation. The results of this method pro-

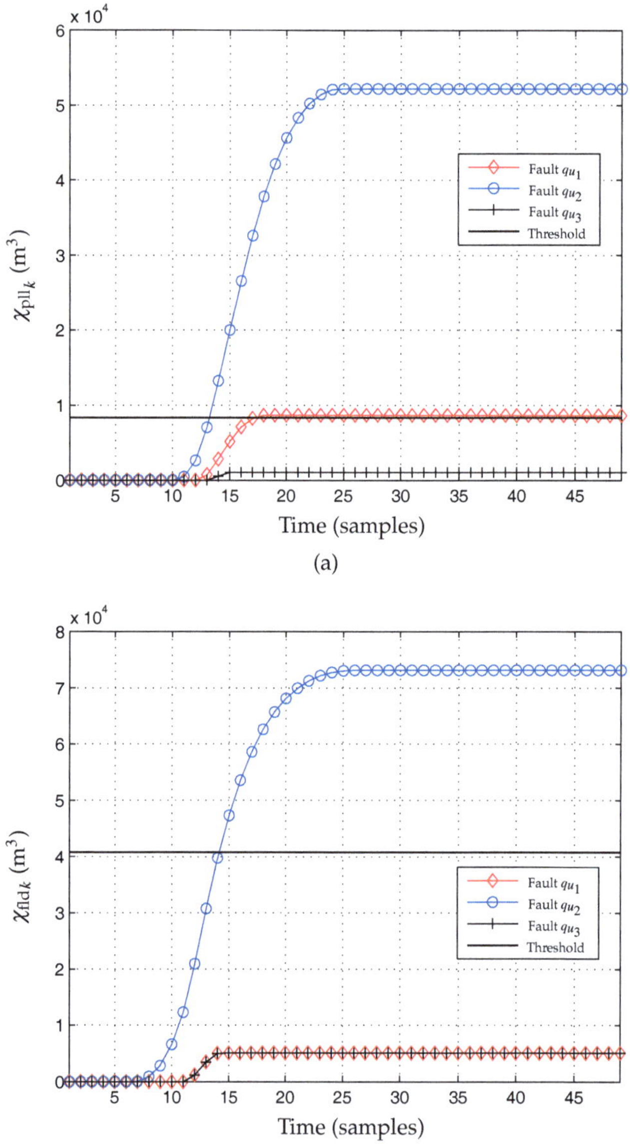

Figure 8.7 Results for minimum volumes in the case with actuator cancellation: (a) pollution, and (b) flooding

vide the performance limits of the system, taking into account all of the feasible solutions and how they degrade after a fault has occurred. The latter result allows to evaluate the admissibility of a given AFC through the use

Table 8.3 Partially damaged actuators: admissibility of AFCs when pollution is considered

Fault location	Operating range	$\chi_{\text{pll}_{H_p}}$ (m^3)	Admissibility evaluation
No fault	—	1050	—
Fault in q_{u_1}	0–20%	5200	Admissible
Fault in q_{u_1}	0–50%	2300	Admissible
Fault in q_{u_2}	0–20%	34000	Inadmissible
Fault in q_{u_2}	0–50%	15700	Inadmissible

Table 8.4 Partially damaged actuators: admissibility of AFCs when flooding is considered

Fault location	Operating range	$\chi_{\text{fld}_{H_p}}$ (m^3)	Admissibility evaluation
No fault	—	5100	—
Fault in q_{u_1}	0–20%	5100	Uncertain
Fault in q_{u_1}	0–50%	5100	Uncertain
Fault in q_{u_2}	0–20%	50000	Uncertain
Fault in q_{u_2}	0–50%	26100	Inadmissible

of a degradation criterion that is established beforehand. Other techniques can also be used to evaluate AFC admissibility. These techniques are based on set computation using zonotopes (briefly explained in Section 8.3.2), subpavings and other approximations.

The proposed method was successfully applied to a system inspired by the BTC within the framework of LCMPC. This three-tank catchment (3-TC) contained representative elements of the entire sewer network and considered three of the four control gates that appear in the BTC. For these reasons, the 3-TC is sufficient to show the effectiveness of the proposed approach when applied to a real system. The technique introduced in this chapter was also shown to work with a nonlinear model of the 3-TC given by the following expressions (Ocampo-Martinez *et al.*, 2006a):

$$\begin{aligned} v_{2k+1} &= v_{2k} + \Delta t \left[q_{u_1 k} + d_{2k} - q_{v_{2out} k} \right], \\ v_{3k+1} &= v_{3k} + \Delta t \left[q_{u_2 k} - q_{u_3 k} \right], \\ v_{4k+1} &= v_{4k} + \Delta t \left[q_{u_3 k} + d_{3k} + q_{14k} + q_{24k} - q_{v_{4out} k} \right], \end{aligned} \tag{8.27}$$

where a nonlinear relation between the i-th tank volume v_i and the tank outflow $q_{v_{iout}} = \beta_i \sqrt{v_{ik}}$ was considered. Results reported in Ocampo-Martinez *et al.* (2006a) show the effectiveness of this approach despite the nonlinear nature of the prediction model associated with the MPC controller.

On the other hand, Guerra *et al.* (2007) propose an actuator fault tolerance evaluation which can be used with linear constrained robust MPC (LRMPC)

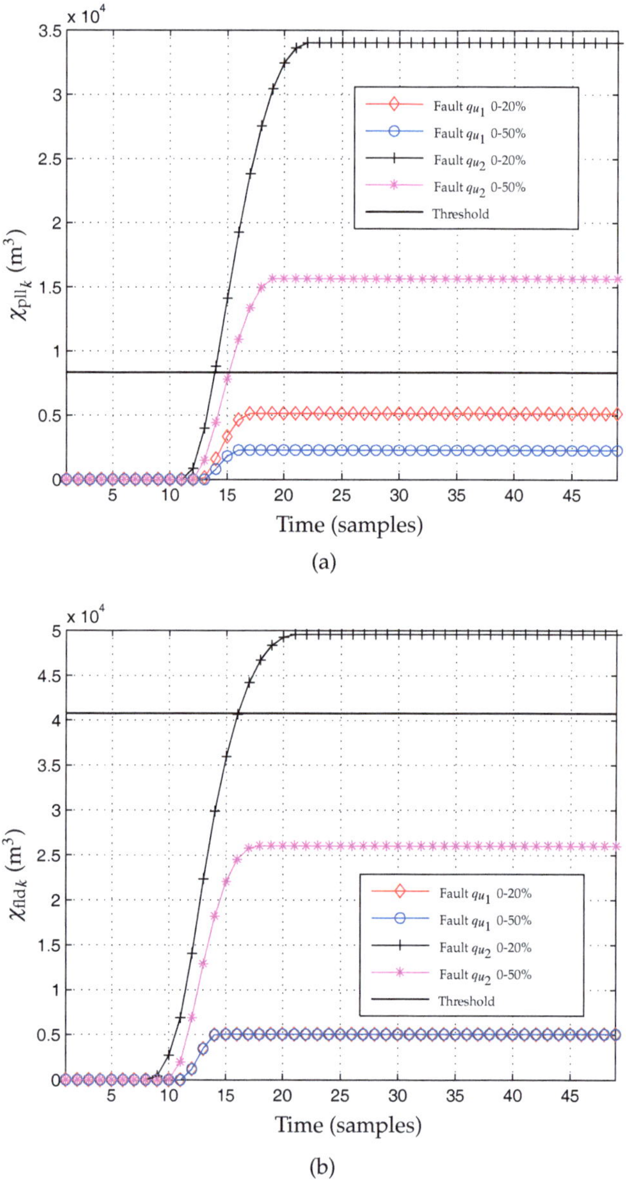

Figure 8.8 Results for minimum volumes in the case with partially damaged actuators: (a) pollution, and (b) flooding

that takes model uncertainty into account. In this case, the problem considers the parameters to be new variables that are refined when the ICSP is stated. In that work, the effectiveness of the admissibility evaluation approach is proven despite the presence of parameter uncertainty.

Part IV
Concluding Remarks

Chapter 9
Concluding Remarks

9.1 Final Discussion

The central idea of this book was to design MPC controllers for sewer networks while considering the nature of the predictive model associated with the controller. Additionally, complementary topics such as fault tolerance have been explored. Figure 9.1 depicts a conceptual scheme that shows the order in which the different topics treated in this book were presented and how they are related.

In general terms, the dynamical models of sewer networks employed in this book were used for control purposes, resulting in linear models that were not very complex but which were accurate enough to represent the dynamical behaviour of the networked system over a prediction time window. The complexity of the whole system depends on the size of the network and so the number of interconnected elements and how they are related. [1] Note that the feedback provided by the measurements sent from the system enables the correction of the prediction model, thus affording a valid representation of the network despite the fact that the nonlinear dynamics are not modelled.

Once the linear model for global control had been defined, some linear MPC controllers were designed and tested. A tuning methodology based on lexicographic minimisers was proposed, and several tests were performed utilising different cost function norms associated with the optimisation problem behind the MPC controller design. On the other hand, a methodology for evaluating the admissibility of an actuator fault configuration was proposed, discussed and tested in the framework of the described LCMPC scheme.

---

[1] According to the definition of "complex", which comes from the Latin *complexus*, an item is complex when it is composed of diverse elements. Note that although "complex" is also used as a synonym for "complicated", strictly speaking, something that is complex does not necessarily have to be complicated.

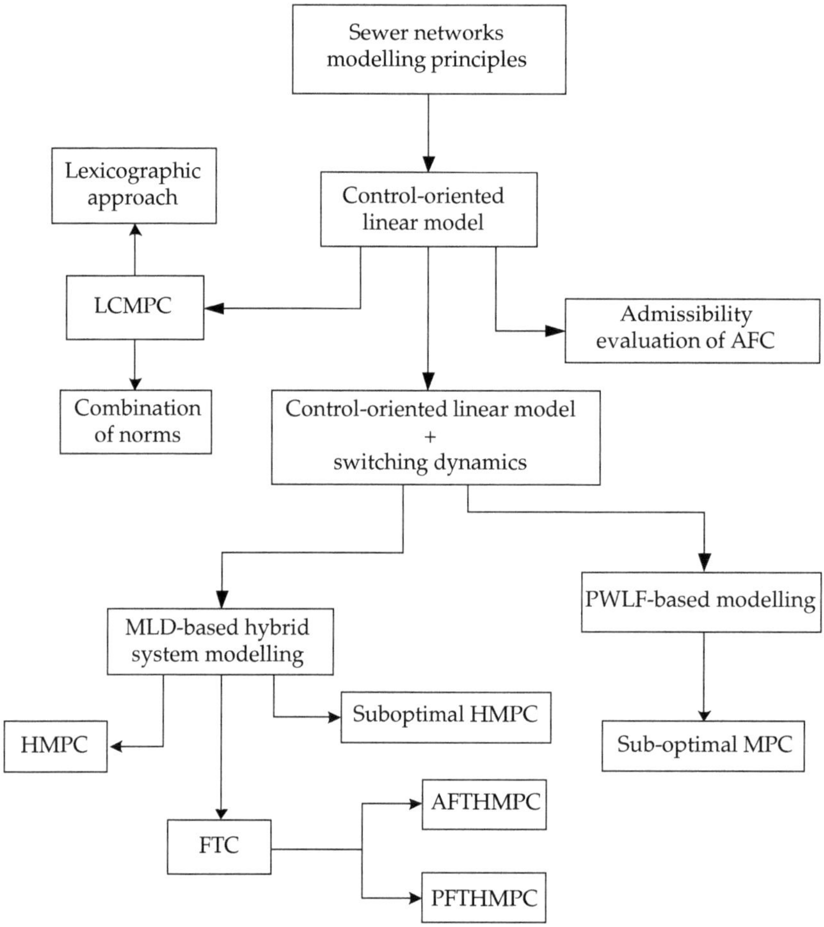

Figure 9.1 Scheme showing the different topics treated in this book

Next, the switching dynamics of the system were considered, which are generated by certain phenomena and/or the behaviours of some constitutive elements. This led to two approaches. On the one hand, hybrid system modelling based on MLD forms was proposed. On the other, a PWLF-based modelling methodology was also described. These two methodologies mainly differ in how Boolean variables are handled during the MLD-based modelling, which implies discrete optimisation problems during the computation of control actions, thus yielding $\mathscr{N}\mathscr{P}$-hard problems (where the computational burden grows exponentionally as a function of the number of Boolean variables), despite the resulting globally optimal solution, because the optimisation problem remains convex. By contrast, the PWLF-based modelling approach implies non-convex optimisation problems, which result in some degree of performance degradation but also a significant reduction in the computational burden. As will be pointed out in the next section, if the MPC controller associated with the PWFL-based model is formulated adequately, a quasiconvex optimisation problem may be stated that yields logarithmic computation time and a bounded suboptimal performance, in contrast to the fully convex optimisation problem.

Following the hybrid system modelling branch in Figure 9.1, the inclusion of fault tolerance mechanisms in the design of an HMPC was also discussed. Considering the intrinsic multi-modal nature of hybrid systems, it seems natural to include faults as new system modes. This implies the availability of a complete architecture for fault diagnosis and isolation, so that information on the fault is made available and can be included as a mode in the hybrid system model and then in the controller design. All of these procedures conform to the AFTHMPC methodology, as explained in Chapter 7.

Some of the principal results obtained, explained and discussed in this book are collected and outlined as follows:

- The lexicographic approach was used as an automatic tuning mechanism for the MPC controller of a sewer network. The application of this technique to such a complex system was motivated by the difficulty involved in determining the appropriate weights for the cost function of a traditional MPC controller due to continuous changes in rain intensity (system disturbances).
- A hybrid modelling methodology for sewer networks was developed. The proposed methodology allows each constitutive element of the network to be represented as a single hybrid system. To obtain the model of the entire system within this framework, all pre-established hybrid submodels are merged appropriately (taking into account the original connections between them), thus avoiding the tedious and complex process of having to model the entire sewer network as a single hybrid system.
- The predictive control of a sewer network was defined by considering its model to be a hybrid system. This allows the computation of the globally optimal solution of the associated optimisation problem, although the sewage system model includes nonlinear dynamics relating to overflows and flooding, which implies the switching of operating modes.
- A suboptimal HMPC design was derived in order to reduce the computational burden of solving the MIP problem behind the MPC controller design. The resulting approach guarantees the feasibility of the optimisation problem and the closed-loop stability of the control scheme. Such a suboptimal HMPC design was tested on the case study, and satisfactory results were obtained in terms of reducing the computational burden as well as the degree of networked system performance degradation.

- An alternative suboptimal MPC approach was proposed. This approach was based on modelling the switching dynamics of the networked system using piecewise linear functions, which avoids the need to use model methodologies that include Boolean variables. When this approach was applied, a considerable reduction was observed in the computation times of the control actions with respect to the corresponding times when a hybrid model was utilised for the same goal. Additionally, the loss of performance seen when using this modelling methodology was comparable to that associated with hybrid modelling when the prediction model of the networked system was obtained.
- The hybrid modelling methodology was employed in order to include actuator faults, which were considered to be changes in the operating ranges of these elements. Moreover, the hybrid nature of the FTC system was taken into account by applying hybrid system modelling and control methodologies. This allowed the three levels of the FTC system to be designed in an integrated manner and its global behaviour to be verified.
- A method for evaluating the fault tolerance of an LCMPC closed-loop scheme under the influence of actuator faults was proposed. This method employs constraint satisfaction to check whether a certain actuator fault configuration fulfils the problem constraints, or whether the corresponding control objectives are fulfilled despite the presence of such actuator faults. This method of evaluating tolerance avoids the need to solve an optimisation problem to determine whether the control law can deal with the given AFC.

The methodologies and control designs proposed along this book can easily be interpolated to other complex systems of a similar nature, such as drinking water transportation networks, systems of irrigation canals, and many others. In other words, the research described here can be viewed in a more general context too.

9.2 Possible Directions for Future Research

The research already performed on sewer network control has indicated several ways in which techniques and methodologies used for the smart management of such systems can be studied and enhanced. Just as in other cases where a real case study leads to the creation or modification of theoretical approaches to solving a given problem, sewer networks are complex systems that have many different issues that must be addressed, ranging from their design (in the topological and planning sense) to the formidable challenges of control and performance analysis. However, sewer networks share many features with other networks associated with the urban water cycle. Thus, any theoretical development or methodological improvement in this area can potentially also be applied to drinking water transportation networks, systems of irrigation canals, water collection networks, *etc.*

This book has explored the suitability of applying MPC to large-scale networks that convey wastewaters. Nevertheless, this is a wide field of study, and some of its areas still await detailed study. A variety of topics could and should be explored in the future, such as:

- *Suitable modelling methodologies for large-scale networked systems with switching dynamics.* This is an open research topic, as method development depends greatly on the case under study (Sastry and Raja-Rao, 2001; Parmar *et al.*, 2007; Döll *et al.*, 2008).
- *Forecasting disturbances and their inclusion in predictive models for MPC controller design in sewer network management.* When designing MPC controllers for models that are subject to measured disturbances, it is important to define the model of these disturbances so that they can be *predicted* over the prediction horizon. Previous studies have shown that they can be considered to be constant or known, and that the system performance does not significantly change (Gelormino and Ricker, 1994). However, forecasting disturbance behaviour properly would lead to more realistic MPC strategies. Rain intensity – the main disturbance in unitary sewer networks – is very difficult to predict. Hence, the adequacies of models based on radar, time series, *etc.* should be considered in the management of these networked systems.
- *Tuning methodologies for MPC controllers applied to large-scale networked systems.* There are different control objectives depending on the type of network associated with the urban water cycle. For each individual network, the definition of its management objectives depends on many features, such as its nature, its function, its size, *political* interests, *etc.* The main control objectives of sewer networks were outlined in Chapter 4, which also discussed their prioritisation. The way of implementing that prioritisation is to employ a tuning methodology for the MPC controller; *e.g.*, that the order of the control objectives must be respected. This book utilised a weight-based approach for most of the MPC controller designs described in it, and the lexicographic approach for LCMPC controllers in particular. However, a variety of tuning methodologies for multi-objective optimisation problems are reported in the literature (Wojsznis *et al.*, 2003; Ahmad and Wahid, 2007; Logist *et al.*, 2009). Evaluating the suitability of each of these methodologies for application to sewer network control problems is an open research field of great interest.
- *Formulation of quasiconvex optimisation problems in large-scale networked systems with switching dynamics.* When the PWLF-based modelling approach is used to obtain the predictive model of a sewer network, the resulting optimisation problem is non-convex. However, the monotonic and continuous nature of such PWL functions allows the formulation of a quasiconvex optimisation problem. Although the proper formalisation of such problems is already underway in the field of sewer network

modelling, several topics in this field should be analysed further. This would be expected to lead to the generalisation of promising results to any large-state networked system with switching dynamics.
- *Fault-tolerance mechanisms for large-scale hybrid systems.* When systems are modelled using hybrid system approaches, the analysis of FTC architectures is convenient in situations where faults are common and can significantly affect the system performance (see Yang *et al.*, 2010, and references therein). Indeed, this is an important issue in sewer network management, where a fault in either a sensor or an actuator can break the control loop and the entire system can easily collapse. As discussed, this book deals with particular models of actuator faults, but the analysis of other fault models in actuators (as well as in sensors and/or tanks and pipes; *e.g.*, leaks) should be formalised properly and studied in depth.

References

Abdel-Geliel M, Badreddin E, Gambier A (2006) Application of model predictive control for fault tolerant system using dynamic safety margin. In: Proc Am Control Conf, Minneapolis, MN, USA, 14–16 June 2006

Aggelogiannaki E, Sarimveis H (2006) Multiobjective constrained MPC with simultaneous closed-loop identification. Int J Adapt Control Signal Process 20(4):145–173

Ahmad A, Wahid A (2007) Application of model predictive control (MPC) tuning strategy in multivariable control of distillation column. Reaktor 11(2):66–67

Akridge A, Carty D (2006) Real time control implementation within a combined sewer system: an operational perspective. In: Proc World Environ Water Resour Congr, Omaha, NE, USA, 21–26 May 2006

Alamo T, Bravo J, Camacho E (2005) Guaranteed state estimation by zonotopes. Automatica 41(6):1035–1043

Alessio A, Bemporad A (2009) A survey on explicit model predictive control. In: Nonlinear model predictive control: towards new challenging applications (Lecture Notes in Control and Information Sciences, vol 384). Springer, Berlin, pp 345–369

Alwan H, Agarwal A (2009) A survey on fault tolerant routing techniques in wireless sensor networks. In: Proc IEEE Int Conf Sensor Technol Appl, Los Alamitos, CA, USA, 18–23 June 2009, pp 366–371

Ames A, Sastry S (2005) Characterization of zeno behavior in hybrid systems using homological methods. In: Proc Am Control Conf, Portland, OR, USA, 8–10 June 2005, pp 1160–1165

Angeli D, Sontag E (2003) Monotone control systems. IEEE Trans Automat Control 48(10):1684–1698

Attouche S, Hayat S, Staroswiecki M (2001) An efficient algorithm for the design of fault tolerant multi-sensor system. In: Proc IEEE Conf Decision Control, Orlando, FL, USA, 4–7 Dec 2001, 2:1891–1892

Baguenard X (2005) Propagation de contraintes sur les intervalles application à l'étalonnage des robots (Ph.D. thesis). Université d'Angers, Angers

Ballester J, Martí J, Salamero M (1998) Control de compuertas de derivación en la red de alcantarillado de Barcelona. Ing Agua 5(4):37–46

Bazaraa M, Sherali H, Shetty C (2006) Nonlinear programming: theory and algorithms, 3rd edn. Wiley, Hoboken

Bech J, Codina B, Lorente J (2006) Forecasting weather radar propagation condition. Meteorol Atmos Phys 96:229–243

Beck R, Saenger S, Richert F, Bollig A, Neiss K, Scholt T, Noreikat K, Abel D (2005) Model predictive control of a parallel hybrid vehicle drivetrain. In: Proc IEEE Conf Decision Control, Seville, Spain, 12–15 Dec 2005, pp 2670–2675

Bemporad A (2004) Efficient conversion of mixed logical dynamical systems into an equivalent piecewise affine form. IEEE Trans Automat Control 49(5):832–838
Bemporad A (2006) Hybrid Toolbox. http://www.dii.unisi.it/hybrid/toolbox
Bemporad A, Giorgetti N (2006) Logic-based solution methods for optimal control of hybrid systems. IEEE Trans Automat Control 51(6):963–976
Bemporad A, Morari M (1999) Control of systems integrating logic, dynamics, and constraints. Automatica 35(3):407–427
Bemporad A, Borrelli F, Morari M (2002a) On the optimal control law for linear discrete time hybrid systems. In: Tomlin C, Greenstreet M (eds) Hybrid systems: computation and control (Lecture Notes in Computer Science, vol 228). Springer, Berlin, pp 105–119
Bemporad A, Morari M, Dua V, Pistikopoulos E (2002b) The explicit linear quadratic regulator for constrained systems. Automatica 38(1):3–20
Benali M, Hammache A, Aubé F, Dipama J, Cantave R (2007) Dynamic multiobjective optimization of large-scale industrial production systems: an emerging strategy. Int J Energy Res 31(12):1202–1225
Bennett S, Patton R, Daley S (1999) Sensor fault-tolerant control of a rail traction drive. Control Eng Pract 7:217–225
Benosman M (2010) A survey of some recent results on nonlinear fault tolerant control. Math Probl Eng 2010:1–26
Benosman M, Lum K (2008) Passive actuators' fault tolerant control for affine nonlinear systems. In: Proc IFAC World Congr, Seoul, South Korea, 6–11 July 2008
Bertsekas D (2005) Dynamic programming and suboptimal control: a survey from ADP to MPC. Eur J Control 11(4-5):310–334
Bixby R (2002) Solving real-world linear programs: a decade and more of progress. Oper Res 50(1):3–15
Bixby R, Fenelon M, Gu Z, Rothberg E, Wunderling R (2000) MIP: theory and practice – closing the gap. In: Powell M, Scholtes S (eds) System modelling and optimization: methods, theory and applications. Kluwer, Dordrecht, pp 19–49
Blanke M (1999) Fault-tolerant control systems. In: New trends in advanced control. Springer, Heidelberg
Blanke M, Kinnaert M, Lunze J, Staroswiecki M (2006) Diagnosis and fault-tolerant control, 2nd edn. Springer, Berlin
Blondel V, Tsitsiklis J (2000) A survey of computational complexity results in systems and control. Automatica 36:1249–1274
Bordóns C (2000) Control predictivo: metodología, tecnología y nuevas perspectivas (tech rep). Departamento de Ingeniería de Sistemas y Automática, Universidad de Sevilla, Aguadulce, Almería
Borrelli F (2003) Constrained optimal control of linear and hybrid systems (Lecture Notes in Control and Information Sciences). Springer, Heidelberg
Borrelli F, Baotić M, Bemporad A, Morari M (2005a) Dynamic programming for constrained optimal control of discrete-time linear hybrid systems. Automatica 41:1709–1721
Borrelli F, Keviczky T, Balas GJ, Stewart G, Fregene K, Godbole D (2005b) Hybrid decentralized control of large scale systems. In: Morari M, Thiele L (eds) Hybrid systems: computation and control (Lecture Notes of Computer Science, vol 3414). Springer, Berlin, pp 168–183
Boyd S, Vandenberghe L (2004) Convex optimization. Cambridge University Press, Cambridge
Branicky M, Zhang G (2000) Solving hybrid control problems: level sets and behavioral programming. In: Proc Am Control Conf, Chicago, IL, USA, 28–30 June 2000
Branicky MS, Borkar VS, Mitter SK (1998) A unified framework for hybrid control: model and optimal control theory. IEEE Trans Automat Control 43(1):31–45
Branke J, Deb K, Miettinen K (eds) (2008) Multiobjective optimization: interactive and evolutionary approaches (Lecture Notes in Computer Science). Springer, Berlin

Brdys M, Grochowski M, Gminski T, Konarczak K, Drewa M (2008) Hierarchical predictive control of integrated wastewater treatment systems. Control Eng Pract 16(6):751–767

Bringi V, Chandrasekar V (2001) Polarimetric Doppler weather radar: principles and applications. Cambridge University Press, Cambridge

Camacho E, Bordons C (2004) Model predictive control, 2nd edn. Springer, London

Campo P, Morari M (1986) ∞-Norm formulation of model predictive control problems. In: Proc Am Control Conf, Seattle, WA, USA, 18–20 June 2006, pp 339–343

Caraman S, Sbarciog M, Barbu M (2007) Predictive control of a wastewater treatment process. Int J Comput Commun Control 2(2):132–142

Cardei M, Yang S, Wu J (2007) Algorithms for fault-tolerant topology in heterogeneous wireless sensor networks. IEEE Trans Parallel Distr 19:545–558

Cassandras C, Lafortune S, Olsder G (1995) Introduction to the modelling, control and optimisation of discrete event systems. In: Isidori A (ed) Trends in control. Springer, Berlin

Cembrano G, Quevedo J (1999) Optimization in water networks. In: Powell R, Hindi K (eds) Computing and control for the water industry. Research Studies, Baldock

Cembrano G, Quevedo J, Salamero M, Puig V, Figueras J, Martí J (2004) Optimal control of urban drainage systems: a case study. Control Eng Pract 12(1):1–9

Cen L, Xi Y (2009) Aggregation-based model predictive control of urban combined sewer networks. In: Proc 7th Asian Control Conf (ASCC), Hong Kong, China, 27–29 Aug 2009

Chan P, Tam C (2005) Performance and application of a multi-wavelength, ground-based microwave radiometer in rain nowcasting. In: 9th Symp Integrated Observing and Assimilation Systems for Atmosphere, Oceans, and Land Surface, San Diego, CA, USA, 9–13 Jan 2005

Chao HL, Chang CL (2008) A fault-tolerant routing protocol in wireless sensor networks. Int J Sensor Networks 3(1):66–73

Chen J, Patton R, Chen Z (1998) An LMI approach to fault-tolerant control of uncertain systems. In: Proc Joint IEEE Conf Decision and Control (CDC), IEEE Int Symp on Computational Intelligence in Robotics and Automation (CIRA), and IEEE Int Symp on Intelligent Systems and Semiotics (ISAS), Westshore, FL, USA, 16–18 Dec 1998, 1:175–180

Chinchuluun A, Pardalos P (2007) A survey of recent developments in multiobjective optimization. Ann Oper Res 154(1):29–50

CLABSA (2007) Clavegueram de Barcelona S.A. http://www.clabsa.es/

Combastel C (2003) A state bounding observer based on zonotopes. In: Proc Eur Control Conf, Cambridge, UK, 1–4 Sept 2003

Conn AR, Gould N, Sartenaer A, Toint PL (1995) Convergence properties of minimization algorithms for convex constraints using a structured trust region. SIAM J Optimiz 6(4):1059–1086

Crossley A (2005) Accurate and efficient numerical solutions for the Saint-Venant equations of open channel flow (Ph.D. thesis). Faculty of Engineering, University of Nottingham, Nottingham

Cugueró P, Puig V, Saludes J, Escobet T (2002) A class of uncertain linear interval models for which a set based robust simulation can be reduced to few pointwise simulations. In: Proc IEEE Conf Decision Control, Las Vegas, NV, USA, 10–13 Dec 2002, 2:1862–1863

Dardinier-Maron V, Hamelin F, Noura H (1999) A fault-tolerant control design against major actuator failures: application to a three-tank system. In: Proc IEEE Conf Decision Control, Phoenix, AZ, USA, 7–10 Dec 1999, 4:3569–3574

De Schutter B (1999) Optimal control of a class of linear hybrid systems with saturation. In: Proc IEEE Conf Decision Control, Phoenix, AZ, USA, 7–10 Dec 1999, pp 3978–3983

Dellana S, West D (2009) Predictive modeling for wastewater applications: linear and nonlinear approaches. Environ Modell Softw 24(1):96–106

Deshpande A, Patwardhan S, Narasimhan S (2009) Intelligent state estimation for fault tolerant nonlinear predictive control. J Process Contr 19:187–204
Diao Y, Passino K (2001) Stable fault-tolerant adaptive fuzzy/neural control for a turbine engine. IEEE Trans Contr Syst Technol 9:494–509
Diao Y, Passino K (2002) Intelligent fault-tolerant control using adaptive and learning methods. Control Eng Pract 10:494–509
Di Loreto M, Dao M, Jaulin L, Lafay J, Loiseau J (2007) Applied interval computation: a new approach for time-delays systems analysis. In: Chiasson J, Loiseau J (eds) Applications of time delay systems (Lecture Notes in Control and Information Sciences, vol 352). Springer, Berlin, pp 175–197
Döll P, Berkhoff K, Bormann H, Fohrer N, Gerten D, Hagemann S, Krol M (2008) Advances and visions in large-scale hydrological modelling. Adv Geosci 18:51–61
Duchesne S, Mailhot A, Dequidt E, Villeneuve J (2001) Mathematical modeling of sewers under surcharge for real time control of combined sewer overflows. Urban Water 3:241–252
Duchesne S, Mailhot A, Villeneuve J (2004) Global predictive real-time control of sewers allowing surcharged flows. J Environ Eng 130(5):526–534
Earl M, D'Andrea R (2005) Phase transitions in the multi-vehicle task assignment problem. In: Proc Int Mech Eng Congr Exposit (IMECE), Orlando, FL, USA, 5–11 Nov 2005
ElGhaoui L, Calafiore G (2001) Robust filtering for discrete-time systems with bounded noise and parametric uncertainty. IEEE Trans Automat Control 46(7):1084–1089
Elliott A, Trowsdale S (2007) A review of models for low impact urban stormwater drainage. Environ Modell Softw 22(3):394–405
Ermolin Y (1999) Mathematical modelling for optimized control of Moscow's sewer network. Appl Math Modell 23:543–556
Etherington D, Parkes A (2004) Improving coalition performance by exploiting phase transition behavior (tech rep). Computational Intelligence Research Laboratory, University of Oregon, Eugene
Fiacchini M, Alamo T, Albea C, Camacho E (2007) Adaptive model predictive control of the hybrid dynamics of a fuel cell system. In: Proc IEEE Conf Control Appl, Singapore, 1–3 Oct 2007, pp 1420–1425
Figueras J, Cembrano G, Puig V, Quevedo J, Salamero M, Martí J (2002) Coral off-line: an object-oriented tool for optimal control of sewer networks. In: Proc IEEE Int Symp Comput Aided Control Syst Des, Glasgow, 17–20 Sept 2002, 1:224–229
Findeisen R, Allgoer F (2001) An introduction to nonlinear model predictive control. In: Scherer C, Schumacher J (eds) The impact of optimization in control. Dutch Institute of Systems and Control (DISC), Delft, pp 1–45
Floudas C, Pardalos P (2001) Encyclopedia of optimization. Kluwer, Dordrecht
Gambier A (2008) MPC and PID control based on multi-objective optimization. In: Proc Am Control Conf, Seattle, WA, USA, 11–13 June 2008, pp 4727–4732
Gawthrop P, Ballance D (2005) Virtual actuator control of mechanical systems. In: Proc Int Conf Bond Graph Model Simulat (ICBGM), New Orleans, LA, USA, 23–27 Jan 2005
Gelormino M, Ricker N (1994) Model predictive control of a combined sewer system. Int J Control 59:793–816
Gent I, Walsh T (1994) The SAT phase transition. In: Cohn A (ed) Proceedings of 11th European Conference on Artificial Intelligence. Wiley, Amsterdam
Geyer T, Torrisi F, Morari M (2003) Efficient mode enumeration of compositional hybrid models. Hybrid Syst Comput Control 2623:216–232
Geyer T, Papafotiou G, Morari M (2005) Model predictive control in power electronics: a hybrid systems approach. In: Proc IEEE Conf Decision Control, Seville, Spain, 12–15 Dec 2005, pp 5606–5611
Goioia G, Bombardelli F (2002) Scaling and similarity in rough channel flows. Phys Rev Lett 88(1):1–4

Goodwin G, Seron M, de Doná J (2005) Constrained control and estimation: an optimisation approach. Springer, Berlin

GRAHI (2007) Group of Applied Research on Hydrometeorology. http://www.grahi.upc.es/

Granvilliers L, Benhamou F (2006) Realpaver: an interval solver using constraint satisfaction techniques. ACM Trans Math Softw 32(1):138–156

Guerra P, Ocampo-Martinez C, Puig V (2007) Actuator fault tolerance evaluation of linear constrained robust model predictive control. In: Proc Eur Control Conf, Kos, Greece, 2–5 July 2007

Habibi J, Moshiri B, Sedigh A (2006) Suboptimal contractive predictive control for a class of hybrid systems. In: Proc Int Conf Comput Intell Modell Control Automat, Int Conf Intell Agents Web Technol Int Commerce (CIMCA), Sydney, Australia, 28 Nov–1 Dec 2006, pp 123–128

Heemels W, De Schutter B, Bemporad A (2001) Equivalence of hybrid dynamical systems. Automatica 37(7):1085–1091

Heymann M, Lin F, Meyer G, Resmerita S (2002) Analysis of zeno behaviors in hybrid systems (tech rep). Computer Science Department, Technion, Haifa

Hoblos G, Staroswiecki M, Aitouche A (2000) Optimal design of fault tolerant sensor networks. In: Proc IEEE Int Conf Control Appl, Anchorage, AK, USA, 25–27 Sept 2000, 1:467–472

Huo Y, Ioannou P, Mirmirani M (2001) Fault-tolerant control and reconfiguration for high performance aircraft: review (tech rep 01-11-01). University of Southern California, Los Angeles

Hyvnen E (1992) Constraint reasoning based on interval arithmetic: the tolerance approach. Artif Intell 58:71–112

Ikeda K, Shin S (1995) Fault tolerant decentralized adaptive control systems using backstepping. In: Proc IEEE Conf Decision Control, New Orleans, LA, USA, 13–15 Dec 1995, 3:2340–2345

ILOG (2003a) ILOG CPLEX 9.0 parameters. ILOG Inc., Sunnyvale

ILOG (2003b) ILOG CPLEX 9.1 users manual. ILOG Inc., Sunnyvale

Ingimundarson A, Ocampo-Martinez C, Bemporad A (2007) Suboptimal model predictive control of hybrid systems using mode sequence constraints. In: Proc IEEE Conf Decision Control, New Orleans, LA, USA, 12–14 Dec 2007

Ingimundarson A, Bravo J, Puig V, Alamo T, Guerra P (2008) Robust fault detection using zonotope-based set-membership consistency test. Int J Adapt Control Signal Process 23(4):311–330

Jaulin L, Kieffer M, Braems I, Walter E (2001a) Guaranteed nonlinear estimation using constraint propagation on sets. Int J Control 74(18):1772–1782

Jaulin L, Kieffer M, Didrit O, Walter E (2001b) Applied interval analysis, with examples in parameter and state estimation, robust control and robotics. Springer, London

Jiang J (1994) Design of reconfigurable control systems using eigenstructure assignment. Int J Control 59:395–410

Julius A, Sakar M, Bemporad A, Pappas G (2007) Hybrid model predictive control of induction of *Escherichia coli*. In: Proc IEEE Conf Decision Control, New Orleans, LA, USA, 12–14 Dec 2007, pp 3913–3918

Kalapala V (2005) A compilation of results on phase transitions, scale-invariance (Ph.D. thesis). The University of New Mexico, Albuquerque

Karamouz M, Fallahi M, Nazif S, Farahani M (2009) Long lead rainfall prediction using statistical downscaling and artificial neural network modeling. Sci Iranica (Trans A Civ Eng) 16(2):165–172

Katayama K, Kimijima K, Yamanaka O, Nagaiwa A, Ono Y (2006) Stormwater inflow prediction using radar rainfall data compressed by principal component analysis. Water Pract Technol 1(1):67–79

Kerrigan E, Maciejowski J (2002) Designing model predictive controllers with prioritised constraints and objectives. In: Proc IEEE Conf Comput Aided Control Syst Des, Glasgow, UK, 18–20 Sept 2002, 1:33–38
Kerrigan E, Bemporad A, Mignone D, Morari M, Maciejowski J (2000) Multi-objective prioritisation and reconfiguration for the control of constrained hybrid systems. In: Proc Am Control Conf, Chicago, IL, USA, 28–30 June 2000, pp 1694–1698
Khanna G, Bagchi S, Wu YS (2004) Fault tolerant energy aware data dissemination protocol in sensor networks. In: Proc Int Conf Depend Syst Networks (DSN), Washington, DC, USA, 28 June–1 July 2004
Khn W (1998) Rigorously computed orbits of dynamical systems without the wrapping effect. Computing 61(1):47–67
Kieffer M, Jaulin L, Walter E (2002) Guaranteed recursive non-linear state bounding using interval analysis. Int J Adapt Control Signal Process 16(3):193–218
Kuhn F, Moscibroda T, Wattenhofer R (2006) Fault-tolerant clustering in ad hoc and sensor networks. In: Proc 26th IEEE Int Conf Distrib Comput Syst (ICDCS), Lisboa, Portugal, 4–7 July 2006
Kung C (1992) Optimal model matching control for mimo continuous-time systems. In: Proc Singapore Int Conf Intell Control Instrum, Singapore, 17–21 Feb 1992, 1:42–47
Lahoud SEA, Yde L, Bendsen B (1998) Real-time control of the sewer system of Boulogne Billancourt – a contribution to improving the water quality of the Seine. Water Sci Technol 37(1):327–332
Lancaster P, Rodman L (1995) Algebraic Riccati equations. Oxford University Press, New York
Lazar M (2006) Model predictive control of hybrid systems: Stability and robustness (Ph.D. thesis). Technische Universiteit Eindhoven, Eindhoven
Lazar M, Heemels W (2009) Predictive control of hybrid systems: input-to-state stability results for sub-optimal solutions. Automatica 45(1):180–185
Lazar M, Heemels W, Weiland S, Bemporad A (2006) Stability of hybrid model predictive control. IEEE Trans Automat Control 51(11):1813 – 1818
Lazar M, Heemels W, Roset B, Nijmeijer H, van den Bosch P (2008) Input-to-state stabilizing sub-optimal NMPC with an application to DC-DC converters. Int J Robust Nonlin 18(8):890–904
Liang Y, Liaw D, Lee T (2000) Reliable control of nonlinear systems. IEEE Trans Automat Control 45:706–710
Liao F, Wang J, Yang G (2002) Reliable robust flight tracking control: an LMI approach. IEEE Trans Contr Syst Technol 10:76–89
Lincoln B, Rantzer A (2001) Optimizing linear system switching. In: Proc IEEE Conf Decision Control, Orlando, FL, USA, 4–7 Dec 2001, pp 2063–2068
Ljung L (1999) System identification: theory for the user. Prentice Hall, Englewood Cliffs
Logist F, Erdeghem PV, Impe JV (2009) Efficient deterministic multiple objective optimal control of (bio)chemical processes. Chem Eng Sci 64(11):2527–2538
Lunze J, Lamnabhi-Lagarrigue F (eds) (2009) Handbook of hybrid systems control: theory, tools, applications. Cambridge University Press, Cambridge
Lunze J, Richter JH (2006) Design approach to the generalised virtual actuator for control reconfiguration. Automatisierungstechnik 54(7):353–361
Lunze J, Steffen T (2003) Control reconfiguration by means of a virtual actuator. In: Proc IFAC SAFEPROCESS, Washington, DC, USA, 9–11 June 2003, pp 133–138
Lydoire F, Poignet P (2004) Commande non linaire horizon fuyant via l'arithmtique d'intervalles. In: Conf Int Francophone d'Automatique (CIFA), Douz, Tunisia, 22–24 Nov 2004
Lygeros J, Tomlin C, Sastry S (1999) Controllers for reachability specifications for hybrid systems. Automatica 35(3):349 – 370

Lyshevski S, Dunipace K, Colgren R (1999) Identification and reconfigurable control of multivariable aircraft. In: Proc Am Control Conf, San Diego, CA, USA, 2–4 June 1999, 4:2732–2736

Maciejowski J (1997) Reconfiguring control systems by optimization. In: Proc Eur Control Conf, Brussels, Belgium, 1–4 July 1997

Maciejowski J (1998) The implicit daisy chaining property of constrained predictive control. Appl Math Comput Sci 8:695–711

Maciejowski J (1999) Fault-tolerant aspects of MPC. In: Proc IEE Workshop MPC Techniques Appl, London, UK, 29 April 1999

Maciejowski J (2002) Predictive control with constraints. Prentice Hall, Harlow

Maciejowski J, Jones C (2003) MPC fault-tolerant flight control case study: Flight 1862. In: Proc IFAC SAFEPROCESS, Washington, DC, USA, 9–11 June 2003

Maciejowski J, Lemos J (2001) Predictive methods for FTC. In: Åström K, Albertos P, Blanke M, Isidori A, Schaufelberger W, Sanz R (eds) Control of complex systems. Springer, Berlin, pp 229–240

Maciejowski J, Ramirez W (1993) Controlling systems in the face of faults. In: Proc IEE Colloq Fault Diagnos Control Syst Reconfig, London, UK, 26 May 1999

Mancini E, Marcarelli S, Vasilyev I, Villano U (2008) A grid-aware MIP solver: implementation and case studies. Future Gen Comput Syst 24(2):133–141

Mandapaka P, Krajewski W, Ciach G, Villarini G, Smith J (2009) Estimation of radar-rainfall error spatial correlation. Adv Water Resour 32(7):1020–1030

Marinaki M (1999) A non-linear optimal control approach to central sewer network flow control. Int J Control 72(5):418–429

Marinaki M, Papageorgiou M (1997) Central flow control in sewer networks. J Water Resour Plan Manag 123(5):274–283

Marinaki M, Papageorgiou M (1998) Nonlinear optimal flow control for sewer networks. In: Proc Am Control Conf, Philadelphia, PA, USA, 24–26 June 1998, 2:1289–1293

Marinaki M, Papageorgiou M (2001) Rolling-horizon optimal control of sewer networks. In: Proc IEEE Int Conf Control Appl, Mexico City, Mexico, 5–7 Sept 2001, 1:594–599

Marinaki M, Papageorgiou M (2005) Optimal real-time control of sewer networks. Springer, Secaucus

Marker R, Arora J (2004) Survey of multi-objective optimization methods for engineering. Struct Multidiscipl Optimiz 26(6):369–395

Mayne D, Schroeder W (1997) Robust time-optimal control of constrained linear systems. Automatica 33(12):2103 - 2118

Mayne D, Rawlings J, Rao C, Scokaert P (2000) Constrained model predictive control: stability and optimality. Automatica 36:789–814

Mays L (2004) Urban stormwater management tools. McGraw-Hill, New York

Mendonca L, Vieira S, Sousa J, da Costa JS (2006) Fault accommodation using fuzzy predictive control. In: Proc IEEE Int Conf Fuzzy Systems, Vancouver, BC, Canada, 16–21 July 2006

Michaelides MP, Panayiotou CG (2009) Fault tolerant maximum likelihood event localization in sensor networks using binary data. IEEE Signal Process Lett 16(5):406–409

Miettinen K (1999) Non-linear multiobjective optimization. Kluwer, Norwell

Miksch T, Gambier A, Badreddin E (2008) Real-time implementation of fault-tolerant control using model predictive control. In: Proc 17th IFAC World Congr, Seoul, South Korea, 6–11 July 2008, pp 11,136–11,141

Mitchell V, Mein R, McMahon T (2001) Modelling the urban water cycle. Environ Modell Softw 16:615–629

Monasson R, Zecchina R, Kirkpatrick S, Selman B, Troyansky L (1999) Determining computational complexity from characteristic "phase transitions." Nature 400:133–137

Moore R (1966) Interval analysis. Prentice Hall, Englewood Cliffs

Morari M, Baotic M, Borrelli F (2003) Hybrid systems modeling and control. Eur J Control 9:177–189

Negenborn R, Beccuti A, Demiray T, Leirens S, Damm G, Schutter BD, Morari M (2007) Supervisory hybrid model predictive control for voltage stability of power networks. In: Proc Am Control Conf, New York, USA, 11–13 July 2007, pp 5444–5449

Niemann H, Stoustrup J (2005) Passive fault tolerant control of a double inverted pendulum: a case study. Control Eng Pract 13(8):1047–1059

Nudelman E, Leyton-Brown K, Hoos HH, Devkar A, Shoham Y (2004) Understanding random SAT: beyond the classes-to-variables ratio. In: Wallace M (ed) Proceedings of the Tenth International Conference on Principles and Practice of Constraint Programming (Lecture Notes in Computer Science 3258). Springer, Berlin, pp 438–452

Ocampo-Martinez C (2004) Benchmark definition for fault-tolerant control based on Barcelona sewer network (internal rep). ESAII, Universitat Politècnica de Catalunya (UPC), Terrassa

Ocampo-Martinez C, Puig V, Quevedo J, Ingimundarson A (2005) Fault tolerant model predictive control applied on the Barcelona sewer network. In: Proc IEEE Conf Decision Control (CDC), Eur Control Conf (ECC), Seville, Spain, 12–15 Dec 2005

Ocampo-Martinez C, Puig V, Quevedo J (2006a) Actuator fault tolerance evaluation of constrained nonlinear MPC using constraints satisfaction. In: Proc IFAC SAFEPROCESS, Beijing, China, 30 Aug–1 Sept 2006

Ocampo-Martinez C, Tornil S, Puig V (2006b) Robust fault detection using interval constraints satisfaction and set computations. In: Proc IFAC SAFEPROCESS, Beijing, China, 30 Aug–1 Sept 2006

Padhiyar N, Bhartiya S (2009) Profile control in distributed parameter systems using lexicographic optimization based MPC. J Process Control 19(1):100–109

Papadimitriou C (1994) Computational complexity. Addison-Wesley, Reading

Papageorgiou M (1985) Optimal multireservoir network control by the discrete maximum principle. Water Resour Res 21(12):1824–1830

Parisini T, Sacone S (1998) Fault diagnosis and controller re-configuration: a hybrid approach. In: Proc 1998 IEEE/ISIC/CIRA/ISAS Joint Conf, Gaithersburg, MD, USA, 14–17 Sept 1998, pp 163–168

Parmar G, Prasad R, Mukherjee S (2007) A mixed method for large-scale systems modelling using eigen spectrum analysis and Cauer second form. IETE J Res 53(2):93–102

Patton R (1997) Fault-tolerant control: the 1997 situation. In: Proc IFAC SAFEPROCESS, Hull, UK, 26–28 Aug 1997, pp 1033–1055

Patwardhan S, Manuja S, Narasimhan S, Shah S (2006) From data to diagnosis and control using generalized orthonormal basis filters. Part II: Model predictive and fault tolerant control. J Process Control 16:157–175

Pleau M, Methot F, Lebrun A, Colas A (1996) Minimizing combined sewer overflow in real-time control applications. Water Qual Res J Canada 31(4):775–786

Pleau M, Pelletier G, Colas H, Lavalle P, Bonin R (2001) Global predictive real-time control of Quebec urban community's westerly sewer network. Water Sci Technol 43(7):123–130

Pleau M, Colas H, Lavalle P, Pelletier G, Bonin R (2005) Global optimal real-time control of the Quebec urban drainage system. Environ Modell Softw 20:401–413

Polyak B, Sergey S, Nazin A, Durieu C, Walter E (2004) Ellipsoidal parameter or state estimation under model uncertainty. Automatica 40(7):1171–1179

Prakash J, Narasimhan S, Patwardhan S (2005) Integrating model based fault diagnosis with model predictive control. Ind Eng Chem Res 44:4344–4360

Previdi F, Lovera M, Mambretti S (1999) Identification of the rainfall–runoff relationship in urban drainage networks. Control Eng Practice 7:1489–1504

Puig V, Saludes J, Quevedo J (2003) Worst-case simulation of discrete linear time-invariant interval dynamic systems. Reliable Comput 9(4):251–290

Puig V, Cembrano G, Romera J, Quevedo J, Aznar B, Ramón G, Cabot J (2009) Predictive optimal control of sewer networks using CORAL tool: application to Riera Blanca catchment in Barcelona. Water Sci Technol 60(4):869–878

Qu Z, Ihlefeld C, Yufang J, Saengdeejing A (2001) Robust control of a class of nonlinear uncertain systems: fault tolerance against sensor failures and subsequent self recovery. In: Proc IEEE Conf Decision Control, Orlando, FL, USA, 4–7 Dec 2001, 2:1472–1478
Qu Z, Ihlefeld C, Yufang J, Saengdeejing A (2003) Robust fault-tolerant self-recovering control of nonlinear uncertain systems. Automatica 39:1763–1771
Rawlings J, Mayne D (2009) Model predictive control: theory and design. Nob Hill, Madison
Richter J, Schlage T, Lunze J (2007) Control reconfiguration of a thermofluid process by means of a virtual actuator. IET Control Theory Appl 1(6):1606–1620
Ricker N, Lee J (1995) Nonlinear model predictive control of the Tennessee Eastman challenge process. Comput Chem Eng 19(9):961–981
Sanyal A, Lee T, Leok M, McClamroch N (2008) Global optimal attitude estimation using uncertainty ellipsoids. Syst Control Lett 57:236–245
Sastry G, Raja-Rao G (2001) A simplified CFE method for large-scale systems modelling about $s = 0$ and $s = a$. IETE J Res 47(6):327–332
Schechter M (1987) Polyhedral functions and multiparametric linear programming. J Optimiz Theory Appl 53:269–280
Schilling W, Anderson B, Nyberg U, Aspegren H, Rauch W, Harremes P (1996) Real-time control of wasterwater systems. J Hydraul Resour 34(6):785–797
Schütze M, Butler D, Beck B (2002a) Modelling, simulation and control of urban wastewater systems. Springer, London
Schütze M, To T, Jaumar U, Butler D (2002b) Multi-objective control of urban wastewater systems. In: Proc IFAC World Congr, Barcelona, Spain, 21–27 July 2002
Schütze M, Campisanob A, Colas H, Schilling W, Vanrolleghem P (2004) Real time control of urban wastewater systems: where do we stand today? J Hydrol 299:335–348
Schütze M, Erbe V, Haas U, Scheer M, Weyand M (2008) Sewer system real-time control supported by the m180 guideline document. Urban Water J 5(1):69–78
Scokaert P, Mayne D, Rawlings J (1999) Suboptimal model predictive control (feasibility implies stability). IEEE Trans Automat Control 44(3):648–654
Sempere-Torres D, Corral C, Raso J, Malgrat P (1999) Use of weather radar for combined sewer overflows monitoring and control. J Environ Eng 125(4):372–380
Shen W, Chen X, Pons M, Corriou J (2009) Model predictive control for wastewater treatment process with feedforward compensation. Chem Eng J 155(1-2):161–174
Singh V (1988) Hydrologic systems. Volume I: Rainfall-runoff modeling. Prentice Hall, Englewood Cliffs
Smith K, Austin G (2000) Nowcasting precipitation – a proposal for a way forward. J Hydrol 239:34–45
Staroswiecki M (2003) Actuator faults and the linear quadratic control problem. In: Proc IEEE Conf Decision Control, Hawaii, USA, 9–12 Dec 2003, 1:959–965
Staroswiecki M, Hoblos G, Aitouche A (2004) Sensor network design for fault tolerant estimation. Int J Adapt Control Signal Process 18(1):55–72
Steffen T (2005) Control reconfiguration of dynamical systems (Lecture Notes in Control and Information Sciences, vol 320). Springer, Berlin
Steffen T, Michail K, Dixon R, Zolotas A, RMGoodall (2009) Optimal passive fault tolerant control of a high redundancy actuator. In: Proc IFAC SAFEPROCESS, Barcelona, Spain, 30 June–3 July 2009
Suresh S (2007) Short term rain prediction for sustainability of tanks in the tropic influenced by shadow rains. In: Int Symp Rainfall Rate Radio Wave Propagat, Salem, India, 29–30 Jan 2007
Teel A (2007) Robust hybrid control systems: an overview of some recent results. In: Bonivento C, Isidori A, Marconi L, Rossi C (eds) Advances in control theory and applications, vol 353. Springer, Berlin, pp 279–302

Theilliol D (2003) Contribution à l'étude et au développement des systèmes tolérants aux defaults: diagnostic et accommodation à base de modèles linéaires et au-delà (Habilitation à Diriger des Recharches). Centre de Recherche en Automatique de Nancy, Université Henri Poincaré, Nancy

Theilliol D, Noura H, Ponsart J (2002) Fault diagnosis and accommodation of a three-tanks system based on analytical redundancy. ISA Trans 41:365–382

Thielen J, Boudevillain B, Andrieu H (2000) A radar data based short-term rainfall prediction model for urban areas – a simulation using meso-scale meteorological modelling. J Hydrol 239(4):97–114

Thorndahl S, Willems P (2008) Probabilistic modelling of overflow, surcharge and flooding in urban drainage using the first-order reliability method and parameterization of local rain series. Water Res 42(1-2):455–466

Till J, Engell S, Panek S, Stursberg O (2004) Applied hybrid system optimization: an empirical investigation of complexity. Control Eng Pract 12:1291–1303

Tomlin C, Lygeros J, Sastry S (2000) A game theoretic approach to controller design for hybrid systems. Proc IEEE 88(7):949–970

Tøndel P, Johansen T, Bemporad A (2003) An algorithm for multi-parametric quadratic programming and explicit MPC solutions. Automatica 39:489–497

Torrisi F (2003) Modeling and reach-set computation for analysis and optimal control of discrete hybrid automata (Ph.D. thesis). Automatic Control Laboratory, Swiss Federal Institute of Technology, Zurich

Torrisi F, Bemporad A (2004) HYSDEL – a tool for generating computational hybrid models for analysis and synthesis problems. IEEE Trans Contr Syst Technol 12(2):235–249

Tyler M, Morari M (1999) Propositional logic in control and monitoring problems. Automatica 35:565–582

Vada J, Slupphaug O, Johansen T, Foss B (2001) Linear MPC with optimal prioritized infeasibility handling: application, computational issues and stability. Automatica 37:1835–1843

Vanrolleghem PA, Benedetti L, Meirlaen J (2005) Modelling and real-time control of the integrated urban wastewater system. Environ Modell Softw 20(4):427–442

Velasco-Forero C, Sempere-Torres D, Cassiraga E, Gómez-Hernández J (2009) A non-parametric automatic blending methodology to estimate rainfall fields from rain gauge and radar data. Adv Water Resour 32(7):986–1002

Veres S, Xia H (1998) Dual predictive control for fault tolerant control. In: Proc UKACC Int Conf Control, Swansea, UK, 1–4 Sept 1998, 2:1163–1168

Villarini G, Smith JA, Baeck ML, Sturdevant-Rees P, Krajewski WF (2010) Radar analyses of extreme rainfall and flooding in urban drainage basins. J Hydrol 381(3-4):266–286

Wahlin B (2004) Performance of model predictive control on ASCE Test Canal 1. J Irrig Drain E – ASCE 130:227–238

Waltz D (1975) Understanding line drawings of scenes with shadows. In: Winston P (ed) The psychology of computer vision. McGraw-Hill, New York, pp 19–91

Weyand M (2002) Real-time control in combined sewer systems in Germany: some case studies. Urban Water 4:347–354

Wojsznis W, Gudaz J, Blevins T, Mehta A (2003) Practical approach to tuning MPC. ISA Trans 42(1):149–162

Wojtaszek D, Chinneck J (2009) Faster MIP solutions via new node selection rules. Comput Operat Res 37(9):1544–1556

Xu B, Wikle C, Fox N (2005) A kernel-based spatio-temporal dynamical model for nowcasting radar precipitation. J Am Stat Assoc 100(472):1133–1144

Yang H, Jiang B, Cocquempot V (2010) Fault tolerant control design for hybrid systems (Lecture Notes in Control and Information Sciences). Springer, Berlin

Yuan J, Tilford K, Jiang H, Cluckie I (1999) Real-time urban drainage system modelling using weather radar rainfall data. Phys Chem Earth B 24:915–919

Zaw W, Naing T (2008) Empirical statistical modeling of rainfall prediction over Myanmar. World Acad Sci Eng Technol 46(1):565–568

Zhang Y, Jiang J (2002) Active fault-tolerant control system against partial actuator failures. IEEE Proc Control Theory Appl 149(1):95–104

Zhang Y, Jiang J (2003) Bibliographiccal review on reconfigurable fault-tolerant control systems. In: Proc IFAC SAFEPROCESS, Washington, DC, USA, 9–11 June 2003, pp 265–276

Zhang Y, Jiang J (2008) Bibliographical review on reconfigurable fault-tolerant control systems. Annu Rev Control 32(2):229–252

Zheng T, Wu G, He D (2007) Nonlinear model predictive control based on lexicographic multi-objective genetic algorithm. In: Proc Int Conf Intell Adv Syst (ICIAS), Kuala Lumpur, Malaysia, 25–28 Nov 2007, pp 61–65

Zhou Y (2004) Model predictive control-based fault tolerant system design. In: Proc World Congr Intell Control Automat, Hangzhou, China, 14–18 June 2004

Zhu C, Park C, Lee W, Yun W (2008) Statistical downscaling for multi-model ensemble prediction of summer monsoon rainfall in the Asia-Pacific region using geopotential height field. Adv Atmos Sci 25(5):867–884

Zhu G, Henson M, Megan L (2001) Dynamic modelling and linear model predictive control of gas pipeline networks. J Process Control 11:129–148

Zug M, Faure D, DeBelly B, Phan L (2001) Use of real time control modelling on the urban sewage system of Nancy. Water Sci Technol 44(2–3):261–268

Zumoffen D, Basualdo M, Molina G (2008) Improvements in fault tolerance characteristics for large chemical plants: pulp mill process with model predictive control. Ind Eng Chem Res 47:5482–5500

Index

Other titles published in this series (continued):

Soft Sensors for Monitoring and Control of Industrial Processes
Luigi Fortuna, Salvatore Graziani, Alessandro Rizzo and Maria G. Xibilia

Adaptive Voltage Control in Power Systems
Giuseppe Fusco and Mario Russo

Advanced Control of Industrial Processes
Piotr Tatjewski

Process Control Performance Assessment
Andrzej W. Ordys, Damien Uduehi and Michael A. Johnson (Eds.)

Modelling and Analysis of Hybrid Supervisory Systems
Emilia Villani, Paulo E. Miyagi and Robert Valette

Process Control
Jie Bao and Peter L. Lee

Distributed Embedded Control Systems
Matjaž Colnarič, Domen Verber and Wolfgang A. Halang

Precision Motion Control (2nd Ed.)
Tan Kok Kiong, Lee Tong Heng and Huang Sunan

Optimal Control of Wind Energy Systems
Iulian Munteanu, Antoneta Iuliana Bratcu, Nicolaos-Antonio Cutululis and Emil Ceangă

Identification of Continuous-time Models from Sampled Data
Hugues Garnier and Liuping Wang (Eds.)

Model-based Process Supervision
Arun K. Samantaray and Belkacem Bouamama

Diagnosis of Process Nonlinearities and Valve Stiction
M.A.A. Shoukat Choudhury, Sirish L. Shah, and Nina F. Thornhill

Magnetic Control of Tokamak Plasmas
Marco Ariola and Alfredo Pironti

Real-time Iterative Learning Control
Jian-Xin Xu, Sanjib K. Panda and Tong H. Lee

Deadlock Resolution in Automated Manufacturing Systems
ZhiWu Li and MengChu Zhou

Model Predictive Control Design and Implementation Using MATLAB®
Liuping Wang

Fault-tolerant Flight Control and Guidance Systems
Guillaume Ducard

Predictive Functional Control
Jacques Richalet and Donal O'Donovan

Fault-tolerant Control Systems
Hassan Noura, Didier Theilliol, Jean-Christophe Ponsart and Abbas Chamseddine

Control of Ships and Underwater Vehicles
Khac Duc Do and Jie Pan

Detection and Diagnosis of Stiction in Control Loops
Mohieddine Jelali and Biao Huang

Stochastic Distribution Control System Design
Lei Guo and Hong Wang

Dry Clutch Control for Automotive Applications
Pietro J. Dolcini, Carlos Canudas-de-Wit, and Hubert Béchart

Active Control of Flexible Structures
Alberto Cavallo, Giuseppe De Maria, Ciro Natale and Salvatore Pirozzi

Nonlinear and Adaptive Control Design for Induction Motors
Riccardo Marino, Patrizio Tomei and Cristiano M. Verrelli

Advanced Control and Supervision of Mineral Processing Plants
Daniel Sbárbaro and René del Villar (Eds.)

Active Braking Control Systems Design for Vehicles
Sergio M. Savaresi and Mara Tanelli

MIX
Papier aus verantwortungsvollen Quellen
Paper from responsible sources
FSC® C105338

If you have any concerns about our products,
you can contact us on
ProductSafety@springernature.com

In case Publisher is established outside the EU,
the EU authorized representative is:
Springer Nature Customer Service Center GmbH
Europaplatz 3, 69115 Heidelberg, Germany

Printed by Libri Plureos GmbH
in Hamburg, Germany